Cybersecurity Dictionary for Everyone

1250 Terms Explained in Simple English

Tolga TAVLAS

Cover Design by London Design Company

Illustrations by Mary Amato

Disclaimer

The information contained in this cybersecurity dictionary is intended for educational purposes only. The author has made every effort to ensure that the information is accurate and up-to-date. However, please be advised that:

- **Cybersecurity terms may have different meanings in different contexts.** The definitions and explanations provided herein are based on generally accepted industry standards, but variations may exist depending on the specific field or application.
- **Cybersecurity is constantly evolving.** The terms, definitions, and explanations presented in this dictionary may change over time as new technologies and threats emerge.
- **No warranty of accuracy.** While every effort has been made to ensure accuracy, there is no guarantee that the information contained in this dictionary is 100% accurate or complete.

The author disclaims all liability for any damages arising from the use of the information contained in this dictionary and is not responsible for any negative consequences that may arise from using the information provided in the book. Users are encouraged to consult with cybersecurity professionals for specific advice and guidance.

Contents

To all the cybersecurity pros I've had the privilege of knowing, meeting, and working with – the brilliant minds, tireless workers, and genuinely good people whose dedication protects not just their organizations but society as a whole.

Preface

My journey into the wild world of cybersecurity started back in the late '90s. Armed with a criminology degree and a healthy dose of curiosity, I was ready to take on the world of cybercrime. Back then, hackers were the stuff of movies, not everyday news. However, after two decades in digital banking, I witnessed firsthand the rapid evolution of these threats. Firewalls, antivirus software, intrusion detection systems... it was a never-ending arms race against a new breed of bad guys who were getting smarter and sneakier by the day.

I quickly realized that to understand and combat these threats truly, you need to speak the language of cybersecurity. Now, cybersecurity might sound like a dry subject to some (trust me, it's not!), but the jargon can be a bit overwhelming. It's a world filled with acronyms, technical terms, and enough buzzwords to make your head spin. That's where this dictionary comes in.

Think of this dictionary as your trusty sidekick in the fight against cybercrime. It's packed with clear, concise explanations of all those confusing terms, sprinkled with real-world examples and a dash of humor to keep things lively. Consider it your secret decoder ring for the digital age.

This isn't my first rodeo. I wrote my first book, "Digital Banking Tips," almost a decade ago. While it didn't exactly make me a millionaire, it did something far more valuable: it helped people. It was used in universities, referenced in articles, and even made it onto a few "best of" lists. But the real reward was knowing that I'd made a difference, contributing to the development of digital banking services in different parts of the world for the better. Now, I want to do the same for cybersecurity, and that's why I wrote this book.

The field of cybersecurity is constantly evolving, with new terms emerging faster than you can say "ransomware!" I've done my best to capture the most common and relevant terms in this dictionary, but I'm sure I've missed a few or perhaps haven't explained them as clearly as I'd like. After all, even cybersecurity experts have their off days. Just remember, context is key. Some terms have different meanings in different settings (some even have surprisingly colorful interpretations on the street!), so use your best judgment, and don't be afraid to ask for help if you need it.

So, whether you're a student dipping your toes into the cybersecurity waters, a seasoned IT professional, or just someone who wants to understand what the heck your IT or Security department is talking about, this dictionary is for you. It's for anyone wanting to be prepared, informed, and maybe even chuckle along the way. Because in the world of cybersecurity, knowledge is power, and a little laughter never hurts either!

Introduction

In an era where cyber threats lurk around every digital corner, understanding cybersecurity is no longer a luxury but a necessity. The recent ransomware attacks on critical infrastructure and data breaches impacting millions of users have highlighted the need for greater awareness and understanding of this complex field. It's more important than ever to understand the language of cybersecurity.

"**Cybersecurity Dictionary for Everyone**" is your essential guide to navigating this complex landscape. It's designed to empower you with the knowledge you need to protect yourself, your business, and your data from the ever-evolving threat landscape. Within these pages, you will find:

- **1250 Essential Cybersecurity Terms:** A comprehensive collection of the most critical terms you need to know.
- **Crystal-Clear Definitions:** Explanations in plain, simple English that anyone can understand, regardless of their technical background.
- **Real-World Examples:** Each definition is illustrated with real-world examples, illustrating how these concepts impact your everyday life and the headlines you read.
- **Related Terms:** Connections between concepts are highlighted, helping you better understand the cybersecurity landscape.

Whether you're a concerned person, a business owner, or a cybersecurity professional, this dictionary will equip you with the knowledge you need to navigate the digital world safely. It goes beyond simple definitions, providing clear explanations and real-world examples to help you understand how these terms apply to everyday life and recent events. It aims to empower everyone with the knowledge they need to protect themselves and their organizations from cyber threats.

This dictionary is your guide to understanding cybersecurity terminology. Use it to stay informed about the latest threats, protect your personal information, and make informed decisions about your digital security. By making cybersecurity terminology accessible and understandable, we can all contribute to a safer and more secure digital future.

Access Control

definition: The selective restriction of access to a place or other resource.

explanation: Like a bouncer at a club, access control checks your "ID" (password, fingerprint, etc.) to ensure you're authorized before granting entry.

real-world examples: Passwords, PINs, biometrics, firewalls, and file permissions.

related terms: Authentication, Authorization, Identity and Access Management (IAM)

Access Management

definition: Managing user identities and their permissions within a system.

explanation: It's like a building manager assigning keys and access cards to different employees, giving them access to specific areas based on their roles.

real-world examples: Employee onboarding/offboarding procedures, software for managing user accounts and permissions, and regular access log reviews.

related terms: Access Control, Identity Management, Privilege Management

Access Token

definition: A temporary credential that grants access to a specific resource or system.

explanation: It's like a concert ticket that proves you've paid for admission and allows you to enter the venue for a specific time.

real-world examples: Using your Google account to log into a third-party website, authorizing a mobile app to access your contacts, or logging into your bank account online.

related terms: Authentication, Authorization, OAuth

Account Takeover (ATO)

definition: When an attacker gains unauthorized access to a user account, typically through stolen credentials.

explanation: It's like a thief getting hold of your keys and taking over your house.

real-world examples: A hacker using stolen login details to access a victim's bank account and make unauthorized transactions.

related terms: Credential Stuffing, Phishing, Identity Theft, Fraud Detection

Active Cyber Defense (ACD)

definition: A proactive approach to cybersecurity that involves actively hunting for and neutralizing threats.

explanation: Imagine a security guard who actively patrols a neighborhood, looking for suspicious activity and intervening to prevent crimes before they happen.

real-world examples: Employing decoy systems, using deception technologies, and actively scanning for and patching vulnerabilities.

related terms: Threat Hunting, Cyber Threat Intelligence (CTI), Incident Response

Active Directory

definition: A directory service that stores information about objects or assets on a network.

explanation: It's like a phonebook for a corporate network, storing information about users, computers, printers, and other resources.

real-world examples: Managing user accounts and computer settings, authenticating users, and authorizing resource access.

related terms: LDAP, Domain Controller, Group Policy

Active Reconnaissance

definition: The process of actively collecting information about a target's systems and vulnerabilities.

explanation: It's like chasing a bank before a robbery, looking for weaknesses in security systems or routines.

real-world examples: Scanning networks for open ports, using social engineering techniques to gather information from employees, and researching a company's online presence.

related terms: Penetration Testing, Threat Intelligence, Vulnerability Scanning

Adaptive Authentication

definition: A security mechanism that adjusts authentication requirements based on the risk level of a login attempt.

explanation: It's like a security guard asking for additional identification if you look suspicious.

real-world examples: Requiring additional verification steps for logins from new locations or devices, using risk-based authentication systems, and implementing multi-factor authentication.

related terms: Multi-Factor Authentication (MFA), Risk-Based Authentication (RBA)

Advanced Encryption Standard (AES)

definition: A symmetric encryption algorithm used to protect sensitive data.

explanation: It's like a strong lockbox that only authorized individuals with the correct key can open.

real-world examples: Encrypting files on your computer, securing online transactions, and protecting sensitive government communications.

related terms: Encryption, Cryptography, Data Security

Advanced Persistent Threat (APT)

definition: A sophisticated and stealthy cyberattack often carried out by a nation-state or organized crime group.

explanation: Imagine a group of highly skilled burglars who patiently observe a house, plan their entry, and steal valuables without leaving a trace.

real-world examples: Targeting government agencies, critical infrastructure, or large corporations to steal sensitive information or disrupt operations.

related terms: Cyber Espionage, Nation-State Attack, Targeted Attack

Advanced Threat Protection (ATP)

definition: A set of security solutions intended to detect, prevent, and respond to sophisticated cyber threats.

explanation: It's like having a highly trained security team that can spot and stop advanced attacks before they cause damage.

real-world examples: Using ATP tools to identify and block phishing attempts, malware, and zero-day exploits.

related terms: Threat Detection, Endpoint Protection, Cyber Threat Intelligence (CTI), Incident Response

Adversarial Machine Learning

definition: The practice of tricking or manipulating machine learning models.

explanation: It's like a magician using sleight of hand to fool an audience; in this case, the magician is tricking a machine learning algorithm.

real-world examples: Spammers tricking email filters, fraudsters fooling facial recognition systems, or adversaries manipulating self-driving cars.

related terms: Machine Learning, Artificial Intelligence (AI), Cybersecurity

Adversary Emulation

definition: Simulating real-world attacks to test an organization's defenses.

explanation: It's like a fire drill, but for cyberattacks – you practice responding to a simulated attack to prepare for the real thing.

real-world examples: Conducting red team exercises, using attack simulation tools, and hiring ethical hackers to test security defenses.

related terms: Penetration Testing, Red Teaming, Ethical Hacking

Adware

definition: Software that shows unwanted advertisements on a user's computer or device.

explanation: It's like those annoying telemarketers who call during dinner – they interrupt your activities and try to sell you something you don't want.

real-world examples: Pop-up ads, banner ads, and unwanted toolbars in your web browser.

related terms: Malware, Spyware, Bloatware

Agile Development

definition: A software development methodology that emphasizes flexibility and collaboration.

explanation: It's like building a house one room at a time, getting feedback, and adjusting along the way, rather than building the whole house according to a rigid plan.

real-world examples: Iterative development, continuous integration, and continuous deployment (CI/CD), and frequent new features or updates releases.

related terms: DevOps, Scrum, Kanban

AI-powered Attacks

definition: Cyberattacks that leverage artificial intelligence to enhance their effectiveness.

explanation: It's like using a supercomputer to plan and execute a heist.

real-world examples: AI-driven phishing attacks that create more convincing emails.

related terms: Machine Learning, Cybersecurity, Phishing

Air Gap

definition: A physical security measure isolates a computer or network from external connections.

explanation: It's like keeping your valuables in a locked safe that is not connected to the internet – it's physically impossible for someone to steal them remotely.

real-world examples: Protecting critical infrastructure systems like nuclear power plants or military networks from cyberattacks.

related terms: Physical Security, Network Isolation

Air-Gapped Network

definition: A network physically isolated or disconnected from other networks, including the internet.

explanation: It's like an island that is not connected to any other landmass - it's very difficult to reach from the outside world.

real-world examples: Protecting highly sensitive data or systems that cannot be exposed to any risk of external compromise.

related terms: Air Gap, Network Isolation

Algorithm

definition: A set of instructions or rules for problem-solving or completion of a task.

explanation: It's like a recipe for baking a cake - it tells you the steps to follow and the ingredients to get the desired result.

real-world examples: Encryption algorithms, hashing algorithms, and search algorithms.

related terms: Cryptography, Data Structures, Computer Science

Annualized Loss Expectancy (ALE)

definition: The expected monetary loss for an asset due to a risk over a one-year period.

explanation: It's like estimating how much money you might lose each year if a particular risk occurs.

real-world examples: Calculating the ALE for data breaches to determine how much money a company might lose annually due to such incidents.

related terms: Risk Management, Risk Assessment, Cost Analysis

Annualized Rate of Occurrence (ARO)

definition: The estimated frequency with which a specific threat or risk is expected to occur within a year.

explanation: It's like predicting how often you might face a particular problem each year.

real-world examples: Estimating that a certain type of cyberattack might happen three times per year.

related terms: Risk Assessment, Probability Analysis, Risk Management

Anonymizer

definition: A tool or service that masks a user's IP address and other identifying information to protect their identity online.

explanation: It's like wearing a disguise to move around unnoticed in a crowded place.

real-world examples: Using Tor or VPN services to browse the internet anonymously.

related terms: VPN (Virtual Private Network), Proxy Server, Privacy Tools

Anti-Forensics

definition: Techniques used to hinder or impede forensic investigations.

explanation: It's like covering your tracks after committing a crime – you try to erase any evidence linking you to the scene.

real-world examples: Encrypting data, deleting logs, and using rootkits to hide malicious software.

related terms: Forensics, Digital Forensics, Data Recovery

Anti-money Laundering (AML)

definition: Measures, regulations, and procedures to prevent criminals from disguising illegally obtained funds as legitimate income.

explanation: It's like a financial detective ensuring money coming into the bank isn't from criminal activities.

real-world examples: Banks using software to monitor and report suspicious transactions.

related terms: Know Your Customer (KYC), Compliance, Financial Crime

Anti-Tamper Technology

definition: Hardware or software mechanisms designed to prevent unauthorized modifications to a system or device.

explanation: It's like a tamper-proof seal on a product – it's designed to break if someone tries to open it without authorization.

real-world examples: Secure boot, hardware security modules (HSMs), and secure firmware.

related terms: Hardware Security, Software Security, Physical Security

Antivirus

definition: Software designed to detect and remove malicious software from a computer or device.

explanation: It's like a security guard who patrols your computer, looking for and removing any unwanted intruders.

real-world examples: Detecting and removing viruses, worms, trojans, and other types of malware.

related terms: Malware, Cybersecurity, Endpoint Security

API Abuse

definition: Exploitation of vulnerabilities in application programming interfaces (APIs).

explanation: It's like breaking into a house through a poorly secured window.

real-world examples: Attackers using APIs to steal data or disrupt services.

related terms: API Security, Web Application Security, Exploits

Application Control

definition: Software that restricts which applications can run on a system.

explanation: Like a picky eater, application control only allows specific programs to execute on your computer, blocking anything not on the approved list.

real-world examples: Antivirus software that blocks malicious programs, or enterprise tools that prevent employees from running unauthorized software.

related terms: Whitelisting, Blacklisting, Endpoint Security

Application Firewall

definition: A firewall that focuses on protecting specific applications or services.

explanation: Think of it as a bodyguard for your favorite app, shielding it from malicious traffic and attacks.

real-world examples: Web Application Firewall (WAF) that protects web applications from attacks like SQL injection and cross-site scripting (XSS).

related terms: Firewall, Web Application Security, Intrusion Prevention System (IPS)

Application Layer

definition: The seventh layer in the OSI model, responsible for application-to-application communication.

explanation: It's like the top floor of a building where different departments (applications) interact with each other.

real-world examples: HTTP, FTP, SMTP, DNS protocols operate at the application layer.

related terms: OSI Model, Network Protocols, TCP/IP

Application Programming Interface (API)

definition: A set of rules and specifications that allows communication among software applications.

explanation: It's like a waiter taking your order and delivering it to the kitchen, then bringing your food back to you. APIs facilitate communication between different software components.

real-world examples: Google Maps API, Twitter API, Facebook API.

related terms: Software Development, Web Services, Integration

Application Programming Interface (API) Gateway

definition: A software component that acts as an entry point for API requests.

explanation: It's like a receptionist who directs calls to the appropriate department – the API gateway routes API requests to the correct backend services.

real-world examples: Managing and securing access to APIs, enforcing authentication and authorization policies, and collecting analytics data.

related terms: API Management, Microservices, Cloud Computing

Application Programming Interface Security (API Security)

definition: Measures to protect APIs from threats and ensure secure data exchange.

explanation: It's like putting security measures in place for how software applications talk to each other.

real-world examples: Implement authentication and authorization for API endpoints and use rate limiting to prevent abuse.

related terms: API Management, Web Application Security, Oauth, Secure Coding Practices

Application Security

definition: The practice of securing software applications from threats and vulnerabilities.

explanation: It's like installing an alarm system in your house to protect it from burglars. Application security involves measures to prevent unauthorized access, data breaches, and other attacks.

real-world examples: Secure Coding Practices, Vulnerability Scanning, Penetration Testing

related terms: Software Security, Secure Coding, Code Review, DevSecOps

Application Whitelisting

definition: A security approach that allows only running approved applications on a system.

explanation: It's like a VIP guest list – only those on the list are allowed into the party (your computer).

real-world examples: Enterprise security solutions that restrict software usage to a pre-approved list.

related terms: Application Control, Blacklisting

Armored Virus

definition: A type of malware that uses various techniques to protect itself from analysis and detection.

explanation: Think of it as a virus wearing a suit of armor, making it difficult for antivirus software to recognize and remove it.

real-world examples: Polymorphic viruses, metamorphic viruses, and viruses that use encryption or obfuscation.

related terms: Malware, Antivirus Evasion, Polymorphic Code

ARP Spoofing (Poisoning)

definition: A type of attack that sends fake ARP messages to connect an attacker's MAC address with the IP address of a legitimate computer or server.

explanation: It's like impersonating someone else to gain access to their belongings. In ARP spoofing, the attacker masquerades as a trusted device to intercept network traffic.

real-world examples: Man-in-the-middle attacks, data theft, and unauthorized access to network resources.

related terms: ARP Cache Poisoning, Network Security, Cyber Attack

Artificial Intelligence (AI)

definition: A branch of computer science that enables machines to simulate human intelligence.

explanation: AI can learn, reason, solve problems, perceive its environment, and even understand language, much like a human would.

real-world examples: Self-driving cars, facial recognition software, recommendation algorithms.

related terms: Machine Learning, Deep Learning, Neural Networks

Asset

definition: Any resource that has value to an organization, including hardware, software, data, and people.

explanation: In cybersecurity, assets are the things we aim to protect. They are the "crown jewels" of an organization.

real-world examples: Servers, laptops, databases, intellectual property, and employees.

related terms: Risk Management, Asset Management, Cybersecurity

Asset Inventory

definition: A comprehensive list of all assets within an organization, including hardware, software, and data.

explanation: It's like keeping a detailed list of everything valuable you own.

real-world examples: Maintaining an up-to-date inventory of a company's computers, servers, and software applications.

related terms: Asset Management, Configuration Management, IT Asset Management (ITAM), Security Posture

Asset Management

definition: The process of tracking and managing an organization's assets.

explanation: It's like keeping an inventory of all valuables to ensure nothing is lost or stolen.

real-world examples: Using software to monitor IT assets and their lifecycles.

related terms: IT Asset Management, Inventory Control, Cybersecurity

Asymmetric Encryption

definition: A cryptographic system that encrypts and decrypts data using a pair of keys (public and private).

explanation: It's like having a locked mailbox with a slot. Anyone can put a message in (encrypt with the public key), but only the owner with the private key can open it (decrypt).

real-world examples: Secure email communication (PGP), SSL/TLS for website security, Bitcoin transactions.

related terms: Encryption, Public Key Cryptography, Private Key

Asynchronous Encryption

definition: Another term for asymmetric encryption, where encryption and decryption use different keys.

explanation: Same concept as asymmetric encryption, emphasizing that the sender and receiver do not need to share the same secret key.

real-world examples: Secure email communication (PGP), SSL/TLS for website security, Bitcoin transactions.

related terms: Encryption, Public Key Cryptography, Private Key

ATM Jackpotting

definition: A type of attack where criminals use malware or hardware to make ATMs dispense cash.

explanation: It's like cracking open a safe without leaving a trace.

real-world examples: Hackers installing malware on ATMs to withdraw large sums of money.

related terms: Cybercrime, Malware, Physical Security

Attack Attribution

definition: The process of identifying the source or perpetrator of a cyberattack.

explanation: It's like a detective investigation: following the digital clues to determine who is responsible for a cybercrime.

real-world examples: Analyzing malware code, tracking IP addresses, and examining attack patterns.

related terms: Cyber Threat Intelligence (CTI), Digital Forensics, Incident Response

Attack Graph

definition: A visual representation of the possible paths an attacker could take to compromise a system.

explanation: Like a map of all the possible routes a burglar could take to break into a house, an attack graph helps visualize potential attack paths.

real-world examples: Identifying and prioritizing vulnerabilities to defend against potential attacks.

related terms: Threat Modeling, Vulnerability Assessment, Penetration Testing

Attack Surface

definition: The sum of all the vulnerabilities in a system that could be exploited by an attacker.

explanation: It's like the number of windows and doors in a house: the more there are, the more potential entry points for a burglar.

real-world examples: Open ports, unpatched software, default passwords, and misconfigured settings.

related terms: Vulnerability Assessment, Penetration Testing, Security Hardening

Attack Vector

definition: The method or pathway that an attacker uses to gain unauthorized access to a system.

explanation: It's the tool a burglar uses to break in: a crowbar, lockpick, or even a phishing email.

real-world examples: Malware, phishing emails, zero-day vulnerabilities, social engineering.

related terms: Vulnerability, Exploit, Threat

Audit Evidence

definition: The information collected and used to support the findings and conclusions in an audit.

explanation: It's like the clues and documents a detective gathers to solve a case.

real-world examples: Financial statements, records, and documentation reviewed during a financial audit.

related terms: Internal Audit, External Audit, Compliance

Audit Trail

definition: A step-by-step record showing the history of transactions or changes in a system.

explanation: It's like a breadcrumb trail that helps track where changes were made and by whom.

real-world examples: Logs in financial systems that record each transaction made.

related terms: Logs, Monitoring, Compliance

Authentication

definition: The process of verifying the identity of a user, device, or system.

explanation: It's like showing your ID to a security guard to prove you are who you say you are.

real-world examples: Passwords, biometrics (fingerprint, facial recognition), security tokens.

related terms: Access Control, Authorization, Identity Management

Authentication Factor

definition: A piece of information that is used for the verification of a user's identity.

explanation: It's like a puzzle piece that, when combined with other pieces, confirms you're who you say you are.

real-world examples: Passwords, PINs, fingerprints, security tokens, and one-time codes.

related terms: Authentication, Multi-Factor Authentication (MFA)

Authentication Header (AH)

definition: A protocol used in IPSec that provides authentication and integrity for data packets.

explanation: It's like a tamper-proof seal on a letter, ensuring it hasn't been altered in transit and comes from the sender you expect.

real-world examples: Used in VPNs and other secure network communications.

related terms: IPSec, Encryption, Network Security

Authorization

definition: The process of granting or denying access to a resource based on a user's identity and permissions.

explanation: Once you're in the club (authenticated), authorization determines which VIP areas you can access.

real-world examples: Assigning roles and permissions to employees in a company's network, setting file permissions on your computer.

related terms: Access Control, Identity, and Access Management (IAM)

Automated Threat Intelligence

definition: The use of software to collect, analyze, and interpret threat data from various sources.

explanation: It's like having a team of security analysts working around the clock to gather and analyze information about potential threats, but without the coffee breaks.

real-world examples: Threat intelligence platforms that aggregate data from multiple sources and provide actionable insights to security teams.

related terms: Threat Intelligence, Security Information and Event Management (SIEM)

Availability

definition: The accessibility and usability of a system, network, or data when needed.

explanation: It's like having your car always ready to go when you need it, without any breakdowns or flat tires.

real-world examples: Ensuring that websites are always online, critical systems are running, and data backups are available for recovery.

related terms: Reliability, Uptime, Disaster Recovery

Availability Zone

definition: A physical location within a cloud computing region designed to be isolated from failures in other zones.

explanation: It's like having multiple power generators for a hospital – if one fails, the others can keep the lights on.

real-world examples: Cloud providers use availability zones to ensure high availability and fault tolerance for their services.

related terms: Cloud Computing, Disaster Recovery, High Availability

Backdoor

definition: A hidden method for bypassing normal authentication or encryption in a computer system.

explanation: It's like a secret passage in a castle that allows someone to bypass the main gate and guards.

real-world examples: Malicious backdoors can be installed by hackers to gain unauthorized access to a system, while legitimate backdoors may be used for troubleshooting or maintenance.

related terms: Malware, Remote Access Trojan (RAT), Vulnerability

Backscatter

definition: Email messages that are bounced back to the original sender after being rejected by a spam filter.

explanation: It's like a boomerang – you throw it out, and it comes back to you. A backscatter can be a sign that your email address has been spoofed by spammers.

real-world examples: Receiving emails that you didn't send or noticing that your emails are being blocked by spam filters.

related terms: Spam, Email Spoofing, Email Security

Backtrace

definition: A report or record of the sequence of function calls that were active at a certain point in a program's execution, typically used for debugging purposes.

explanation: It's like retracing your steps to find out where you went wrong after getting lost.

real-world examples: Developers using backtraces to identify the cause of a crash in a software application.

related terms: Debugging, Stack Trace, Error Handling, Crash Report

Bait and Switch

definition: A deceptive technique where an attacker substitutes a malicious file or link for a legitimate one.

explanation: It's like ordering a pizza and getting a pineapple instead - it's not what you expected, and it could have negative consequences.

real-world examples: A user clicks on a link they believe will take them to a safe website, but they are in fact, redirected to a malicious one.

related terms: Malware, Phishing, Social Engineering

Baiting

definition: A social engineering tactic that relies on the victim's curiosity or greed to lure them into a trap.

explanation: It's like leaving a wallet full of cash on the sidewalk, hoping someone will pick it up and fall victim to a scam.

real-world examples: Leaving a USB drive with malware on it in a public place or sending a phishing email with a too-good-to-be-true offer.

related terms: Social Engineering, Phishing, Malware

Basel III

definition: A global regulatory framework targeted at strengthening the regulation, supervision, and risk management within the banking sector.

explanation: It's like a safety net ensuring banks have enough capital to cover risks and avoid financial crises.

real-world examples: Banks increasing their capital reserves to meet Basel III requirements.

related terms: Risk Management, Regulatory Compliance, Financial Stability

Baseline

definition: A snapshot of a system's security configuration at a specific point in time.

explanation: It's like a before photo in a home renovation project – it shows you what the system looked like before any changes were made.

real-world examples: Used to compare against future scans to identify unauthorized changes or potential security vulnerabilities.

related terms: Security Assessment, Configuration Management, Change Management

Bastion Host

definition: A heavily fortified server located on the edge of a network, acting as a gateway to internal resources.

explanation: It's like a heavily guarded fortress that protects a kingdom from invaders.

real-world examples: Often used to provide secure remote access to internal networks or to host publicly accessible services.

related terms: DMZ, Firewall, Security Hardening

Beaconing

definition: The act of a compromised system sending regular signals to a remote server controlled by an attacker.

explanation: It's like a homing beacon that allows the attacker to track and communicate with the compromised system.

real-world examples: Malware often uses beaconing to receive instructions from a command-and-control (C2) server.

related terms: Malware, Command and Control (C2), Botnet

Business Email Compromise (BEC)

definition: A type of scam targeting businesses to steal money or sensitive information through email fraud.

explanation: It's like a con artist impersonating a company executive to trick employees.

real-world examples: Fraudsters sending fake invoices to company employees for payment.

related terms: Phishing, Social Engineering, Fraud

Behavior Anomaly Detection

definition: A security technique that identifies unusual or suspicious activity by comparing it to a baseline of normal behavior.

explanation: It's like a security camera that alerts you when someone is doing something they shouldn't be, like trying to break into a locked room.

real-world examples: Detecting unusual login patterns, data transfers, or file access attempts.

related terms: Intrusion Detection System (IDS), Security Information and Event Management (SIEM)

Behavioral Analytics

definition: The use of technology to analyze patterns in user behavior to detect unusual activities.

explanation: It's like monitoring how people usually act to spot any unusual behavior that might indicate a security threat.

real-world examples: Detecting unusual login times that suggest someone else might be using an account or identifying strange spending patterns in a bank account.

related terms: UEBA, Anomaly Detection, Insider Threat Detection, Machine Learning

Behavior-Based Detection / Security

definition: A security approach detects threats by focusing on monitoring and analyzing user and entity behavior.

explanation: It's like a detective profiling a suspect to predict their next move.

real-world examples: Analyzing user activity logs to identify suspicious patterns or deviations from normal behavior.

related terms: User and Entity Behavior Analytics (UEBA), Security Information and Event Management (SIEM)

Biometric Authentication

definition: A security method that uses unique biological characteristics to verify a person's identity.

explanation: It's like using your fingerprint or face as a password - it's something unique to you that is difficult to forge.

real-world examples: Fingerprint scanning, facial recognition, iris scanning, voice recognition.

related terms: Authentication, Multi-Factor Authentication (MFA)

Biometrics

definition: The measurement and analysis of unique physical or behavioral characteristics for identification and authentication purposes.

explanation: It's like a personalized signature - your fingerprints, face, iris, or voice are unique to you and can be used to verify your identity.

real-world examples: Fingerprint scanning, facial recognition, iris scanning, voice recognition.

related terms: Authentication, Biometric Authentication

Birthday Attack

definition: A type of cryptographic attack that exploits the probability of collisions in hash functions.

explanation: It's like trying to find two people in a room who share the same birthday - it's more likely than you might think.

real-world examples: Used to find weaknesses in cryptographic algorithms or to forge digital signatures.

related terms: Cryptography, Hash Function, Collision Attack

Birthday Paradox

definition: A statistical phenomenon where it's surprisingly likely that two or more people in a group share the same birthday.

explanation: Imagine it's like a lottery where you don't pick your own number. With enough people in the room, the odds are surprisingly good that two will have the same birthday!

real-world examples: This concept is used in some cryptographic attacks to find collisions in hash functions.

related terms: Hash Functions, Cryptography, Collision Attacks

Black Box Testing

definition: A testing method where the tester does not have any knowledge of the internal workings of the system being tested.

explanation: Think of it like trying to figure out how a toy works by just playing with it - you don't know what's inside, but you can observe its behavior.

real-world examples: Security researchers use this to find vulnerabilities in software without access to its source code.

related terms: Penetration Testing, White Box Testing, Gray Box Testing

Black Hat Hacker

definition: A malicious hacker who exploits vulnerabilities for personal gain or malicious intent.

explanation: They are the "bad guys" of the cybersecurity world, motivated by profit, revenge, or simply the thrill of the challenge.

real-world examples: Cybercriminals who steal financial data, launch ransomware attacks, or deface websites for fun.

related terms: Hacker, White Hat Hacker, Cybercrime

Black Hole

definition: A network routing mechanism that silently discards incoming or outgoing traffic.

explanation: It's like a cosmic vacuum cleaner for network traffic - it sucks in data packets and never spits them out.

real-world examples: Used to mitigate denial-of-service (DoS) attacks or to block unwanted traffic from specific sources.

related terms: Routing, Network Security, DoS Mitigation

BlackEnergy

definition: A malware family known for targeting industrial control systems and critical infrastructure.

explanation: It's a versatile toolkit for hackers, capable of causing widespread disruption and damage to industrial processes.

real-world examples: Used in the 2015 attack on the Ukrainian power grid, causing widespread blackouts.

related terms: Malware, Industrial Control Systems (ICS), Critical Infrastructure

Blacklist

definition: A list of entities (e.g., IP addresses, email addresses, or domains) that are known to be malicious or undesirable.

explanation: It's like a "do not call" list for the internet - if you're on the list, you're not welcome.

real-world examples: Email filters use blacklists to block spam, and web browsers might block access to known malicious websites.

related terms: Whitelist, Spam Filtering, Web Filtering

Blacklisting

definition: The process of adding an entity to a blacklist.

explanation: It's like adding someone's name to the "banned" list at a nightclub - they won't be allowed in anymore.

real-world examples: Blocking an IP address known for sending spam or adding a malicious website to a browser's blacklist.

related terms: Blacklist, Security Policy

Block Cipher

definition: A type of encryption algorithm that operates on fixed-length blocks of data.

explanation: It's like a codebook that scrambles a message one block at a time.

real-world examples: AES, DES, and Blowfish are examples of block ciphers.

related terms: Encryption, Cryptography, Stream Cipher

Blockchain

definition: A distributed, immutable ledger that records transactions across multiple computers.

explanation: It's like a public ledger that everyone can see, but no one can erase or tamper with.

real-world examples: Not only used to secure cryptocurrencies like Bitcoin and Ethereum but also has potential applications in other fields like supply chain management and voting systems.

related terms: Cryptocurrency, Distributed Ledger Technology (DLT), Bitcoin

Blue Team

definition: The defenders in a cybersecurity context, responsible for protecting systems and networks from attacks.

explanation: They are like castle guards, constantly patrolling and defending against attackers.

real-world examples: Cybersecurity analysts, incident responders, and security engineers who monitor, detect, and respond to threats.

related terms: Red Team, Cybersecurity Operations, Incident Response

Bluejacking

definition: An attack that sends unsolicited messages to Bluetooth-enabled devices.

explanation: It's like getting an unwanted message from a stranger on your phone - it's annoying but usually not harmful.

real-world examples: Receiving spam messages or unwanted files over Bluetooth.

related terms: Bluesnarfing, Bluetooth Security

Bluesnarfing

definition: An attack that steals data from a Bluetooth-enabled device without the owner's knowledge or consent.

explanation: It's like a pickpocket stealing your wallet without you noticing.

real-world examples: Stealing contacts, photos, or other personal data from a Bluetooth-enabled phone.

related terms: Bluejacking, Bluetooth Security

Bluetooth Attacks

definition: Exploiting vulnerabilities in Bluetooth technology in order to gain unauthorized access to devices.

explanation: It's like hijacking a phone call by tapping into a Bluetooth connection.

real-world examples: Attackers using Bluetooth to steal data from smartphones.

related terms: Wireless Security, IoT Security, Exploits

Bluetooth Low Energy (BLE) Attacks

definition: Cyberattacks targeting the Bluetooth Low Energy protocol used in many IoT devices.

explanation: It's like sneaking into a secured area using a small gap.

real-world examples: Hackers exploiting BLE vulnerabilities to control smart home devices.

related terms: IoT Security, Bluetooth Security, Cybersecurity

Bluetooth Security

definition: The measures taken to protect Bluetooth-enabled devices and communications from unauthorized access and attacks.

explanation: It's like putting a lock on your Bluetooth connection to prevent unwanted guests from joining the party.

real-world examples: Pairing devices with a PIN, using encryption for data transmission, and keeping Bluetooth turned off when not in use.

related terms: Wireless Security, Pairing, Encryption

Board Risk Oversight

definition: The responsibility of a company's board of directors to oversee risk management policies and practices.

explanation: It's like a ship's captain ensuring the crew is managing the risks of navigation and weather.

real-world examples: Regular board meetings to review risk reports and strategies.

related terms: Governance, Risk Management, Compliance

Bot

definition: A software application that runs automated tasks over the Internet.

explanation: They are like digital robots that can perform repetitive tasks, such as sending emails, clicking on ads, or even launching cyberattacks.

real-world examples: Web crawlers, chatbots, and malicious bots used in DDoS attacks.

related terms: Botnet, Automation, Artificial Intelligence

Bot Herder

definition: An individual or group that controls a botnet.

explanation: They are the puppet masters behind the scenes, directing the actions of the bots.

real-world examples: Cybercriminals who use botnets to launch spam campaigns, DDoS attacks, or steal sensitive information.

related terms: Botnet, Cybercrime, Command and Control

Bot Mitigation

definition: The process of detecting and preventing malicious bot activity.

explanation: It's like pest control for your website or network - you're trying to eliminate the harmful bots while allowing the good ones to operate.

real-world examples: Using CAPTCHAs to distinguish between humans and bots, blocking suspicious IP addresses, and analyzing traffic patterns to identify bot activity.

related terms: Botnet, Web Application Firewall (WAF), Rate Limiting

Botnet

definition: A network of compromised computers or devices that an attacker controls remotely.

explanation: It's like an army of zombie computers, blindly following the orders of their master.

real-world examples: Launching DDoS attacks, sending spam emails, and stealing sensitive information.

related terms: Bot, Bot Herder, Command and Control

Botnet Takedown

definition: The process of disrupting or dismantling a botnet by identifying and neutralizing its command-and-control infrastructure.

explanation: It's like cutting the head off a hydra - if you can take down the botnet's control center, you can disable the entire army.

real-world examples: Law enforcement agencies and cybersecurity researchers collaborating to shut down botnets.

related terms: Botnet, Cybercrime, Malware

Breach

definition: An incident that results in the unauthorized access or disclosure of sensitive data.

explanation: It's like a hole in a fence that allows someone to enter your property without permission.

real-world examples: A hacker gaining access to a company's customer database or a data leak caused by a misconfigured cloud server.

related terms: Data Breach, Data Leak, Cybersecurity Incident

Breach and Attack Simulation (BAS)

definition: A method of testing security defenses by simulating cyberattacks.

explanation: It's like running fake attacks on your system to find weak spots before real attackers do.

real-world examples: Simulating a phishing attack to see how many employees click on a fake email or testing if a hacker can move through your network without being detected.

related terms: Penetration Testing, Red Teaming, Vulnerability Assessment, Cyber Range

Bridge Protocol Data Unit (BPDU) Guard

definition: A network security feature that protects against unauthorized changes to network topology.

explanation: Like a guard dog protecting a bridge, BPDU Guard prevents rogue devices from altering the network structure, ensuring stability and preventing loops.

real-world examples: Used on switch ports to disable them if they receive unexpected BPDU messages, indicating a potential attack or misconfiguration.

related terms: Spanning Tree Protocol (STP), Network Security, Loop Prevention

Bring Your Own Device (BYOD)

definition: A policy that allows employees to use their personal devices for work purposes.

explanation: It's like bringing your lunch from home instead of using the company cafeteria. BYOD offers flexibility but also introduces security risks.

real-world examples: Employees using personal smartphones, laptops, or tablets to access company emails, files, and applications.

related terms: Mobile Device Management (MDM), Endpoint Security, Data Loss Prevention (DLP)

Browser Exploit

definition: A type of attack that takes advantage of vulnerabilities in web browsers.

explanation: It's like finding a weak spot in a castle wall that allows an attacker to bypass the defenses.

real-world examples: Drive-by downloads, cross-site scripting (XSS), and malicious browser extensions.

related terms: Web Application Security, Vulnerability, Malware

Brute Force Attack

definition: A type of cyber attack that uses brute force to gain unauthorized access to a system or account.

explanation: It's like trying every key on a keyring until you find the one that opens the door.

real-world examples: Attempting to guess a password by trying every possible combination of characters.

related terms: Brute Force, Password Cracking, Dictionary Attack

Buffer Overflow

definition: A vulnerability that occurs when a program tries to write data beyond the limits of a buffer. It is writing more data to a memory space (buffer) than it can hold.

explanation: It's like overfilling a glass of water – the excess spills out and can cause a mess. In this case, the "mess" is an opportunity for an attacker to execute malicious code.

real-world examples: Stack overflow, heap overflow.

related terms: Memory Corruption, Vulnerability, Exploit

Buffer Underflow

definition: A vulnerability that occurs when a program reads data from a buffer that has not been fully initialized.

explanation: It's like trying to read the rest of a book after you've reached the last page. The result is unpredictable and potentially dangerous.

real-world examples: Reading sensitive data that was not meant to be accessed, causing crashes or other unexpected behavior.

related terms: Memory Corruption, Vulnerability, Exploit

Buffer Zone

definition: A neutral area between two potentially hostile networks or systems.

explanation: It's like a demilitarized zone (DMZ) between two countries – it acts as a buffer to prevent conflict and provide a layer of security.

real-world examples: A DMZ that separates a company's internal network from the internet.

related terms: DMZ, Network Security, Perimeter Security

Bug Bounty

definition: A program that rewards individuals for finding and reporting software bugs or vulnerabilities.

explanation: It's like a reward for finding a needle in a haystack, except the needle is a security flaw, and the haystack is a complex software application.

real-world examples: Companies like Google, Microsoft, and Apple offer bug bounty programs to incentivize ethical hackers to find and report vulnerabilities before they can be exploited by malicious actors.

related terms: Vulnerability Disclosure, Ethical Hacking, Vulnerability Research

Business Continuity Management (BCM)

definition: Planning and preparing to keep a business running during and after a disaster.

explanation: It's like having a backup plan to keep everything running smoothly even if something goes wrong.

real-world examples: Creating a plan to keep essential services running during a power outage or having a strategy to recover data after a cyberattack.

related terms: Disaster Recovery, Business Continuity Plan (BCP), Risk Management, Incident Response

Business Continuity Plan (BCP)

definition: A plan that outlines how an organization will continue operating during and after a disruptive event.

explanation: It's like a backup plan for when things go wrong – it ensures that critical business functions can continue even if the primary systems or facilities are unavailable.

real-world examples: A plan for how a company will continue operating if its headquarters is destroyed in a natural disaster.

related terms: Disaster Recovery (DR), Risk Management, Incident Response

Business Email Compromise (BEC)

definition: A type of phishing scam that targets businesses by impersonating a trusted individual, such as a CEO or CFO, to initiate fraudulent wire transfers or other unauthorized activities.

explanation: It's like a con artist who impersonates a wealthy investor to trick someone into giving them money.

real-world examples: An email from the "CEO" instructing an employee to wire money to a fraudulent account or an email from the "CFO" requesting sensitive financial information.

related terms: Phishing, Social Engineering, Spear Phishing

Business Impact Analysis (BIA)

definition: A process for identifying critical business functions and assessing the potential impact of those functions.

explanation: It's like a doctor performing a checkup on a patient to assess their overall health and identify any potential risks.

real-world examples: Identifying which business functions are most critical to the company's survival and determining how long they can be disrupted before causing significant harm.

related terms: Risk Management, Business Continuity Planning, Disaster Recovery

Business Logic Attack

definition: An attack that exploits flaws in the design or implementation of a business process or application logic.

explanation: It's like finding a loophole in a contract that allows you to get something for nothing.

real-world examples: An attacker exploits a flaw in an online shopping cart to get a discount on a product or manipulates a bank's transaction system to withdraw more money than they have in their account.

related terms: Application Security, Web Application Security, Vulnerability

Business Logic Vulnerability

definition: A weakness in the design or implementation of a business process or application logic that can be exploited by attackers.

explanation: It's like a design flaw in a building that makes it easier for burglars to break in.

real-world examples: An e-commerce website that doesn't properly validate user input, allowing an attacker to manipulate prices or steal sensitive data.

related terms: Application Security, Web Application Security, Vulnerability

Cache

definition: A temporary storage area that holds frequently and repeatedly accessed data for quick retrieval.

explanation: It's like a pantry that stores your favorite snacks, so you don't have to go to the grocery store every time you're hungry.

real-world examples: Web browsers use caches to store website data, and CPUs have caches to store frequently used instructions.

related terms: Browser Cache, CPU Cache, Web Performance

Cache Poisoning

definition: An attack that manipulates the data stored in a cache to redirect users to malicious websites or inject false information.

explanation: It's like replacing the ingredients in a recipe with something harmful, so anyone who uses that recipe gets sick.

real-world examples: DNS cache poisoning, web cache poisoning.

related terms: DNS Spoofing, Web Application Security

California Consumer Privacy Act (CCPA)

definition: A state statute intended to enhance privacy rights and consumer protection for residents of California, USA.

explanation: It's like giving consumers more control over their personal data and how it's used by companies.

real-world examples: Companies providing transparency reports and opt-out options for data sharing.

related terms: GDPR, Data Privacy, Consumer Rights

Call Back

definition: A security mechanism where a remote system calls back to a pre-authorized number to verify the legitimacy of a connection request.

explanation: It's like a phone call from a friend to confirm they sent you a text message.

real-world examples: Modems and remote access servers often use callback to verify the identity of the caller.

related terms: Authentication, Remote Access Security

Canary Token

definition: A decoy file or resource that is designed to alert security personnel when it's accessed or modified.

explanation: It's like a tripwire alarm - it doesn't prevent an intruder from entering, but it alerts you that someone has crossed the line.

real-world examples: Placing a fake file in a sensitive directory or embedding a hidden image in an email.

related terms: Intrusion Detection, Threat Monitoring, Honeytoken

CAPTCHA

definition: A test designed to distinguish between humans and automated bots by presenting a challenge that is difficult for computers to solve.

explanation: It's like a "prove you're not a robot" test - you might be asked to identify distorted letters or images or solve a simple math problem.

real-world examples: Used to prevent bots from creating fake accounts, submitting spam comments, or buying up tickets to events.

related terms: Bot Mitigation, Web Security, Image Recognition

CAPTCHA Farms

definition: Large-scale operations that employ humans to solve CAPTCHAs for automated bots.

explanation: Imagine a sweatshop where workers are paid to solve CAPTCHAs all day, allowing bots to bypass security measures.

real-world examples: Used by malicious actors to create fake accounts, spread spam, and engage in other automated abuse.

related terms: CAPTCHA, Bot Mitigation, Cybercrime

Card Cloning

definition: The process of creating a duplicate of a credit or debit card.

explanation: It's like making a photocopy of your card but with the ability to make purchases.

real-world examples: Cloning is done by copying the data from the magnetic stripe or chip of a card, often using a skimmer.

related terms: Card Skimming, Identity Theft, Fraud

Card Skimming

definition: The act of stealing credit or debit card information using a hidden electronic device.

explanation: It's like a thief secretly copying your card while you're not looking.

real-world examples: Skimmers are often attached to ATMs or point-of-sale terminals to capture card data.

related terms: Card Cloning, Identity Theft, Fraud

Carding

definition: The trafficking and use of stolen credit card information.

explanation: It's like a black market for stolen credit cards, where criminals buy and sell card data for fraudulent purposes.

real-world examples: Online forums and marketplaces where stolen card data is sold.

related terms: Credit Card Fraud, Cybercrime, Identity Theft

Certificate Authority (CA)

definition: A trusted entity that issues digital certificates to verify the ownership of public keys.

explanation: It's like a passport office that issues official documents to prove your identity.

real-world examples: Companies like DigiCert, GoDaddy, and Let's Encrypt are well-known CAs.

related terms: Public Key Infrastructure (PKI), Digital Certificate, SSL/TLS

Certificate Management

definition: The process of managing digital certificates to ensure secure communications and data integrity.

explanation: It's like keeping track of all your ID cards to make sure they are valid and not expired.

real-world examples: Managing SSL certificates for secure website connections or using digital certificates for email encryption.

related terms: Public Key Infrastructure (PKI), SSL/TLS, Digital Certificates, Encryption

Certificate Pinning

definition: A security mechanism that hardcodes a website's expected certificate into an application.

explanation: It's like having a photo of your friend's ID stored on your phone, so you can compare it to the ID they show you to make sure it's really them.

real-world examples: Prevents man-in-the-middle attacks by ensuring that the website's certificate matches the one the application expects.

related terms: Man-in-the-Middle Attack, SSL/TLS, Public Key Infrastructure (PKI)

Certificate Revocation List (CRL)

definition: A list of digital certificates that have been revoked or marked invalid by the issuing Certificate Authority.

explanation: It's like a list of canceled passports - if your passport is on the list, it's no longer valid.

real-world examples: Browsers and other applications check CRLs to ensure that the certificates they encounter are still valid.

related terms: Certificate Authority (CA), Public Key Infrastructure (PKI), Digital Certificate

Certificate Transparency

definition: An open framework to monitor and audit SSL/TLS certificates.

explanation: It's like a public registry of all issued passports - anyone can check to see if a certificate is legitimate.

real-world examples: Helps to detect and prevent the issuance of fraudulent or malicious certificates.

related terms: SSL/TLS, Certificate Authority (CA), Public Key Infrastructure (PKI)

Chain of Custody

definition: The chronological documentation of the possession, control, transfer, analysis, and disposition of physical or electronic evidence.

explanation: It's like a paper trail that shows who had access to a piece of evidence and what happened to it at each stage.

real-world examples: Used in forensic investigations to ensure the integrity and admissibility of evidence in court.

related terms: Digital Forensics, Evidence, Forensic Investigation

Chain of Trust

definition: A series of interconnected systems and processes that ensure the security and integrity of data and transactions.

explanation: It's like a sequence of locks where each one must be secure to ensure the overall security of the system.

real-world examples: Digital certificates in SSL/TLS ensure secure communication between web browsers and servers.

related terms: Public Key Infrastructure (PKI), Digital Certificates, Trust Anchor

Challenge-Handshake Authentication Protocol (CHAP)

definition: An authentication protocol utilized by Point-to-Point Protocol (PPP) servers to verify and confirm the identity of remote clients.

explanation: It's like a secret handshake that only two people know – it's used to prove their identity to each other.

real-world examples: Used in dial-up connections and some VPNs to authenticate users.

related terms: PPP, Authentication, Remote Access

Challenge-Response Authentication

definition: An authentication scheme where the user is presented with a challenge (e.g., a random number) and must provide the correct response based on a secret value (e.g., a password).

explanation: It's like a security question that only you know the answer to.

real-world examples: Used in two-factor authentication and other security systems to verify user identity.

related terms: Authentication, Two-Factor Authentication (2FA), Security Token

Chatbot Security

definition: The measures taken to protect chatbots from abuse, misuse, and attacks.

explanation: It's like training a chatbot to be polite but firm, so it can handle rude users and protect itself from malicious inputs.

real-world examples: Input validation, content filtering, and anomaly detection for chatbots.

related terms: Artificial Intelligence (AI), Natural Language Processing (NLP), Security

Checksum

definition: A value calculated from a block of digital data to verify its integrity.

explanation: It's like a digital fingerprint that can be used to ensure that the data hasn't been altered or corrupted.

real-world examples: Used to verify the integrity of downloaded files, backups, and other data.

related terms: Data Integrity, Hash Function, Error Detection

Chief Information Security Officer (CISO)

definition: The executive responsible for an organization's information security program.

explanation: They are the "top cop" of cybersecurity, overseeing the company's efforts to protect its information assets.

real-world examples: Developing and implementing security policies, managing risk, and responding to security incidents.

related terms: Cybersecurity, Information Security, Risk Management

Child Sexual Abuse Material (CSAM)

definition: Any visual depiction of sexually explicit conduct involving a minor.

explanation: It's a serious crime that involves the exploitation and abuse of children.

real-world examples: Cybersecurity professionals work to detect and remove CSAM from online platforms.

related terms: Cybercrime, Child Exploitation, Law Enforcement

Children's Online Privacy Protection Act (COPPA)

definition: A U.S. federal law that imposes certain requirements on operators of websites of online services to protect the privacy of children under 13.

explanation: It's like a guardian ensuring children's data isn't collected or shared without parental consent.

real-world examples: Websites requiring parental consent before collecting data from children.

related terms: Data Privacy, Parental Controls, Online Safety

Chopper

definition: A type of malware that modifies or corrupts data on a computer system.

explanation: It's like a vandal who defaces property, but in this case, the property is your data.

real-world examples: Corrupting files, altering system settings, and making data unusable.

related terms: Malware, Data Destruction, Cybercrime

Chosen-Ciphertext Attack

definition: An attack where the attacker can choose ciphertexts to be decrypted and observe the resulting plaintexts, potentially revealing weaknesses in the encryption scheme.

explanation: It's like a spy who intercepts encrypted messages and tries to figure out their meaning by experimenting with different decryption keys.

real-world examples: Used to break weak encryption algorithms or to discover vulnerabilities in cryptographic systems.

related terms: Cryptography, Encryption, Ciphertext

Chosen-Plaintext Attack

definition: A type of attack where the attacker can choose the plaintext to be encrypted and observe the resulting ciphertext.

explanation: It's like a codebreaker who knows some of the words in a message and tries to deduce the rest of the code.

real-world examples: Used to find weaknesses in encryption algorithms.

related terms: Cryptography, Encryption, Plaintext

CIA Triad

definition: A model that defines the three core principles of information security: Confidentiality, Integrity, and Availability.

explanation: It's like a three-legged stool – if one leg is missing, the stool will fall over. Similarly, if one of the CIA principles is compromised, information security is at risk.

real-world examples: Confidentiality: Protecting sensitive data from unauthorized access; Integrity: Ensuring that data is accurate and reliable; Availability: Ensuring that authorized users have access to data when they need it.

related terms: Information Security, Cybersecurity, Risk Management

Cipher

definition: An algorithm for performing encryption or decryption.

explanation: A set of instructions that scrambles (encrypts) or unscrambles (decrypts) data, often using a secret key.

real-world examples: Caesar cipher, AES, RSA.

related terms: Encryption, Cryptography, Algorithm

Cipher Block Chaining (CBC)

definition: A mode of operation for block ciphers that adds an extra layer of complexity by XORing each plaintext block with the previous ciphertext block before encryption.

explanation: Like a chain reaction, each block's encryption depends on the previous block's output. This helps to enhance security and prevent patterns from emerging in the ciphertext.

real-world examples: Commonly used in data encryption standards like AES.

related terms: Block Cipher, Encryption Mode, Initialization Vector (IV)

Cipher Suite

definition: A set of algorithms that define how a secure connection (e.g., HTTPS) is established.

explanation: It's like a recipe for a secure connection, listing the ingredients (algorithms) and the steps to follow.

real-world examples: Choosing a cipher suite involves selecting algorithms for key exchange, authentication, encryption, and message authentication.

related terms: TLS/SSL, Encryption, Public Key Infrastructure (PKI)

Click Fraud

definition: A type of online fraud where malicious actors repeatedly click on pay-per-click (PPC) ads to generate revenue for themselves or deplete a competitor's advertising budget.

explanation: Imagine someone clicking on a competitor's ads all day to drain their budget – that's click fraud.

real-world examples: Fake clicks generated by bots or click farms.

related terms: Ad Fraud, Pay-Per-Click (PPC) Advertising, Online Fraud

Clickjacking

definition: A web attack that when users are tricked into clicking on hidden elements on a webpage.

explanation: It's like putting invisible tape over a button, so when you think you're clicking on one thing, you're clicking on something else.

real-world examples: Attackers use it to steal sensitive information, spread malware, or perform unauthorized actions.

related terms: UI Redressing, Web Application Security, Cross-Site Request Forgery (CSRF)

Client-Side Attack

definition: An attack that targets vulnerabilities in client-side software (e.g., web browsers, email clients).

explanation: It's like a Trojan horse - the attacker disguises their malicious code within seemingly innocent content that runs on the victim's device.

real-world examples: Cross-site scripting (XSS), clickjacking, and malicious browser extensions.

related terms: Web Application Security, Malware, Browser Exploit

Client-Side Certificate

definition: A digital certificate installed on a user's device to authenticate them to a server.

explanation: It's like a digital ID card that proves your identity to a website or application.

real-world examples: Used in client authentication schemes, often in addition to traditional username and password logins.

related terms: Public Key Infrastructure (PKI), Authentication, Digital Certificate

Client-Side Validation

definition: The process of validating user input on the client-side (e.g., within a web browser) before it is submitted to the server.

explanation: It's like double-checking your work before handing it in to your teacher.

real-world examples: Used to improve user experience and provide immediate feedback, but it's not a substitute for server-side validation, which is essential for security.

related terms: Input Validation, Web Application Security, Server-Side Validation

Clone Phishing

definition: A phishing attack where scammers create a fake website that looks nearly identical to a legitimate one.

explanation: It's like creating a fake driver's license that looks just like the real thing.

real-world examples: Attackers use clone phishing to trick users into entering their login credentials or other sensitive information on the fake website.

related terms: Phishing, Social Engineering, Spoofing

Cloud Access Security Broker (CASB)

definition: Security software that sits between an organization's on-premises infrastructure and cloud providers to enforce security policies.

explanation: It's like a security checkpoint at the border of your cloud environment, ensuring that only authorized traffic and users can enter or leave.

real-world examples: Monitoring user activity in the cloud, enforcing data loss prevention (DLP) policies, and providing threat protection.

Cloud Application Security

definition: The practice of securing applications that run in the cloud.

explanation: It's like installing security cameras and alarms in your virtual house in the cloud.

real-world examples: Implementing security controls to protect cloud-based applications from unauthorized access, data breaches, and other threats.

related terms: Cloud Security, Software as a Service (SaaS), Platform as a Service (PaaS)

Cloud Computing Security

definition: The set of technologies, policies, and procedures designed to protect cloud environments and the data stored within them.

explanation: It's like a comprehensive security plan for a virtual city in the cloud, covering everything from physical infrastructure to data encryption to access control.

real-world examples: Implementing firewalls, intrusion detection systems, encryption, and access controls in the cloud.

related terms: Cloud Security, Infrastructure as a Service (IaaS), Cloud Service Provider (CSP)

Cloud Data Protection

definition: The process of safeguarding data stored in the cloud from unauthorized access, theft, or loss.

explanation: It's like putting your valuables in a safe deposit box, but instead of a bank, you're trusting a cloud provider to keep your data secure.

real-world examples: Encrypting data stored in the cloud, implementing access controls, and using cloud backup solutions.

related terms: Cloud Security, Data Security, Encryption

Cloud Encryption

definition: The practice of encrypting data before it's stored in the cloud.

explanation: It's like scrambling a message before sending it through the mail, so only the intended recipient can read it.

real-world examples: Encrypting files before uploading them to cloud storage services, using encryption keys managed by the cloud provider or the customer.

related terms: Cloud Security, Data Security, Encryption

Cloud Malware

definition: Malicious software that specifically targets cloud environments.

explanation: It's like a virus that's designed to spread through the cloud, infecting virtual machines and stealing data.

real-world examples: Malware that exploits vulnerabilities in cloud infrastructure, cloud applications, or cloud-based services.

related terms: Cloud Security, Malware, Virtual Machine (VM) Escape

Cloud Security

definition: The broad set of technologies, policies, and procedures used to protect cloud computing environments.

explanation: It's like a comprehensive security system for a virtual city in the cloud, encompassing everything from physical infrastructure to data protection to access control.

real-world examples: Implementing firewalls, intrusion detection systems, encryption, access controls, and other security measures in the cloud.

related terms: Cloud Computing, Cybersecurity, Cloud Service Provider (CSP)

Cloud Security Alliance (CSA) Framework

definition: A non-profit organization that promotes best practices for cloud security.

explanation: It's like a neighborhood watch group for the cloud, providing guidance and resources to help organizations secure their cloud environments.

real-world examples: Developing industry standards, certifications, and training programs for cloud security.

related terms: Cloud Security, Cybersecurity, Security Framework

Cloud Security Posture Management (CSPM)

definition: Tools and processes used to ensure cloud environments adhere to security best practices and compliance requirements.

explanation: It's like having a checklist to make sure your cloud setup is secure and follows all the rules.

real-world examples: Scanning cloud configurations for vulnerabilities and ensuring compliance with standards like the General Data Protection Regulation (GDPR) or HIPAA.

related terms: Cloud Security, Compliance Management, Risk Assessment, Configuration Management

Cloud Workload Protection Platform (CWPP)

definition: A security solution that protects workloads running in the cloud.

explanation: It's like a security blanket for your cloud applications, shielding them from threats and vulnerabilities.

real-world examples: Monitoring and protecting virtual machines, containers, and serverless functions in the cloud.

related terms: Cloud Security, Cloud Workload Security, Cloud-Native Security

Cloud-based Attacks

definition: Cyberattacks that target cloud computing environments.

explanation: It's like targeting a storage warehouse instead of individual homes.

real-world examples: Data breaches involving cloud storage providers.

related terms: Cloud Security, Data Breach, Cybersecurity

Control Objectives for Information and Related Technologies (COBIT)

definition: A framework for developing, implementing, monitoring, and improving IT governance and management practices.

explanation: It's like a comprehensive guide to managing IT processes and ensuring they align with business goals.

real-world examples: Companies using COBIT to standardize IT governance and ensure compliance.

related terms: IT Governance, Risk Management, Compliance

Code Injection

definition: An attack that involves injecting malicious code into a computer program.

explanation: It's like adding poison to a recipe - the code alters the program's intended behavior, often with harmful consequences.

real-world examples: SQL injection, cross-site scripting (XSS), and command injection.

related terms: Web Application Security, Vulnerability, Exploit

Code of Conduct

definition: A set of rules outlining the social norms, ethical standards, and responsibilities of individuals within an organization.

explanation: It's like the rules of a game, ensuring everyone knows what behavior is acceptable.

real-world examples: Employee handbooks that outline expectations for behavior and ethics.

related terms: Ethics, Compliance, Corporate Governance

Code Signing

definition: The process of digitally signing executable code to verify its authenticity and integrity.

explanation: It's like a notary public's seal on a document, verifying that it hasn't been tampered with and that it comes from the source it claims to be from.

real-world examples: Used to verify the authenticity of software updates, drivers, and other executable files.

related terms: Digital Signature, Public Key Infrastructure (PKI), Software Security

Cold Boot Attack

definition: An attack that exploits the residual data in a computer's memory after it has been powered off.

explanation: It's like a thief stealing information from a discarded notebook that still contains sensitive notes.

real-world examples: Attackers can use special tools to extract data from a computer's RAM even after it has been shut down.

related terms: Memory Forensics, Encryption, Data Remanence

Cold Site

definition: A backup facility with basic infrastructure but no equipment or data pre-installed.

explanation: Imagine a spare office space with power and internet, but you need to bring your own furniture and computers to make it usable.

real-world examples: Used for disaster recovery when a primary site is unavailable.

related terms: Disaster Recovery, Hot Site, Warm Site

Collision Attack

definition: An attack that exploits the possibility of two different inputs producing the same hash value.

explanation: It's like finding two different books with the same ISBN number. It shouldn't happen, but if it does, it can compromise the integrity of the system.

real-world examples: Used to forge digital signatures or find weaknesses in cryptographic hash functions.

related terms: Hash Function, Cryptography, Birthday Attack

Command and Control (C2)

definition: The infrastructure used by cybercriminals to maintain communication with compromised systems within a target network.

explanation: It's like a control room where attackers issue commands and receive data from the infected machines.

real-world examples: Botnet operators using C2 servers to manage and update malware on compromised systems.

related terms: Botnet, Malware, Cyberattack

Command Injection

definition: An attack that injects arbitrary commands into a vulnerable application.

explanation: It's like sneaking your own instructions into someone else's recipe, causing unexpected and potentially harmful results.

real-world examples: An attacker might inject commands into a web application's search bar to execute malicious code on the server.

related terms: Injection Attack, Web Application Security, Vulnerability

Common Platform Enumeration (CPE)

definition: A standardized method for naming and identifying IT products and platforms.

explanation: It's like a universal product code (UPC) for software and hardware, providing a consistent way to refer to them across different systems.

real-world examples: Used in vulnerability databases and security advisories to identify affected products.

related terms: Vulnerability Management, Software Identification, Hardware Identification

Common Vulnerabilities and Exposures (CVE)

definition: A list of publicly disclosed cybersecurity vulnerabilities.

explanation: It's like a catalog of known security weaknesses in software and hardware.

real-world examples: Each vulnerability is assigned a unique identifier (e.g. , CVE-2023-1234) for easy reference.

related terms: Vulnerability, National Vulnerability Database (NVD), Common Vulnerability Scoring System (CVSS)

Compliance

definition: Adherence to a set of rules, regulations, standards, or specifications.

explanation: It's like following the rules of a game - if you don't, you could face penalties or disqualification.

real-world examples: Complying with industry regulations like HIPAA or PCI DSS or adhering to internal security policies.

related terms: Security Policy, Regulatory Compliance, Audit

Compliance Audit

definition: An independent assessment to determine if an organization's security practices meet regulatory or industry standards.

explanation: It's like a health inspector checking a restaurant's kitchen to ensure it meets food safety regulations.

real-world examples: Evaluating an organization's adherence to HIPAA, PCI DSS, or other compliance frameworks.

related terms: Compliance, Audit, Risk Management

Compliance Enforcement

definition: The process of ensuring compliance to laws, regulations, guidelines, and specifications.

explanation: It's like a referee making sure all players follow the rules of the game.

real-world examples: Regulatory agencies imposing fines for non-compliance.

related terms: Compliance Audit, Regulatory Compliance, Monitoring

Compliance Obligations

definition: The requirements an organization must meet to adhere to laws, regulations, and standards.

explanation: It's like a checklist of rules and regulations that must be followed.

real-world examples: Requirements for data protection under GDPR.

related terms: Regulatory Compliance, Risk Management, Compliance Program

Compliance Officer

definition: An individual responsible for overseeing and managing compliance within an organization.

explanation: It's like a lifeguard ensuring everyone follows the pool rules.

real-world examples: A Chief Compliance Officer (CCO) in a financial institution.

related terms: Compliance Program, Regulatory Compliance, Risk Management

Compliance Program

definition: A set of internal policies and procedures implemented by an organization to comply with laws, regulations, and standards.

explanation: It's like a safety net ensuring all operations are within legal and ethical boundaries.

real-world examples: Corporate compliance programs for anti-bribery and corruption.

related terms: Regulatory Compliance, Compliance Officer, Compliance Training

Compliance Risk Assessment

definition: The process of identifying and evaluating risks related to regulatory compliance.

explanation: It's like a health check-up for a company's adherence to regulations.

real-world examples: Assessing risks related to data privacy regulations.

related terms: Risk Management, Compliance Program, Internal Audit

Compromise Assessment

definition: An investigation to determine the extent and impact of a security breach.

explanation: It's like assessing the damage after a fire – you need to figure out what's been destroyed and what needs to be repaired or replaced.

real-world examples: Analyzing logs, interviewing employees, and conducting forensic investigations to determine how a breach occurred and what data was compromised.

related terms: Incident Response, Digital Forensics, Data Breach

Computer Emergency Response Team (CERT)

definition: A group of experts who respond to computer security incidents.

explanation: They are like the firefighters of the digital world, rushing to the scene to put out fires and minimize damage.

real-world examples: Analyzing and responding to cyberattacks, coordinating incident response efforts, and providing guidance to affected organizations.

related terms: Incident Response, Cybersecurity, Vulnerability Disclosure

Computer Fraud and Abuse Act (CFAA)

definition: A U.S. federal law that prohibits unauthorized access to computers and networks.

explanation: It's like a legal shield protecting against cyber intrusions and data theft.

real-world examples: Prosecuting hackers who illegally access computer systems.

related terms: Cybercrime, Data Protection, Legal Compliance

Computer Security

definition: The protection of computer systems and networks from unauthorized access, theft, damage, or disruption.

explanation: It's like locking your doors and windows to protect your home from burglars.

real-world examples: Implementing firewalls, antivirus software, encryption, and access controls.

related terms: Cybersecurity, Information Security, Network Security

Computer Security Incident Response Team (CSIRT)

definition: A team that handles and responds to computer security incidents.

explanation: They are the first responders to a cyber attack, working to contain the damage and restore normal operations.

real-world examples: Investigating the cause of a security incident, notifying affected parties, and implementing measures to prevent future incidents.

related terms: Incident Response, Cybersecurity, Security Operations Center (SOC)

Confidentiality

definition: The principle of keeping sensitive data private and protected from unauthorized access.

explanation: It's like whispering a secret to a friend and trusting them not to tell anyone else.

real-world examples: Encrypting data, using strong passwords, and implementing access controls to prevent unauthorized disclosure of sensitive information.

related terms: Information Security, Privacy, Encryption

Confidentiality Agreement

definition: A legal contract that protects confidential information from being disclosed without authorization.

explanation: It's like a pinky promise – you agree not to share someone's secret with anyone else.

real-world examples: Used to protect trade secrets, customer data, and other sensitive information.

related terms: Non-Disclosure Agreement (NDA), Intellectual Property, Legal

Configuration Management

definition: The process of maintaining the consistency of a system's performance by keeping its configuration up to date.

explanation: It's like regularly tuning up a car to ensure it runs smoothly.

real-world examples: Using configuration management tools to automate updates and patches.

related terms: IT Management, Cybersecurity, System Administration

Conflict of Interest Policy

definition: Guidelines designed to prevent conflicts of interest within an organization.

explanation: It's like a rulebook ensuring personal interests don't interfere with professional duties.

real-world examples: Policies requiring employees to disclose outside business interests.

related terms: Ethics, Corporate Governance, Compliance

Consumer Privacy Protection Act

definition: Legislation aimed at protecting consumer privacy by regulating the collection and use of personal data.

explanation: It's like a privacy guard ensuring companies handle personal data responsibly.

real-world examples: Requiring businesses to implement data protection measures and provide consumers with data rights.

related terms: Data Privacy, Regulatory Compliance, Consumer Rights

Container Security

definition: The practice of securing containerized applications and their environments.

explanation: It's like building a secure container ship to protect cargo from pirates.

real-world examples: Scanning container images for vulnerabilities, enforcing security policies, and isolating containers from each other.

related terms: Containerization, Cloud Security, Application Security

Containerization

definition: A method of operating system virtualization allowing applications to run in isolated user spaces called containers.

explanation: It's like packing your belongings into separate boxes for a move - each container has everything it needs to run independently.

real-world examples: Used to package and deploy applications in a portable and scalable way.

related terms: Docker, Kubernetes, Cloud Computing

Containment

definition: The process of isolating a compromised system or network to prevent further spread of an attack.

explanation: It's like quarantining a sick person to prevent the spread of a disease.

real-world examples: Disconnecting a compromised computer from the network, blocking malicious traffic with a firewall, or disabling compromised user accounts.

related terms: Incident Response, Security Operations Center (SOC)

Content Disarm and Reconstruction (CDR)

definition: A security technique that removes potentially malicious content from files before they are delivered to users.

explanation: It's like sanitizing a package before opening it, removing any harmful substances or objects.

real-world examples: Stripping active content from email attachments, removing macros from documents, and converting files to a safer format.

related terms: Malware Prevention, Email Security, File Security

Content Filtering

definition: The process that blocks or allows access to specific types of content based on predetermined criteria.

explanation: It's like a parental control setting on a TV that prevents children from watching certain channels.

real-world examples: Blocking websites containing inappropriate material, filtering out spam emails, or preventing the download of certain file types.

related terms: Web Filtering, Email Filtering, Internet Filtering

Content Security Policy (CSP)

definition: A security mechanism that helps prevent cross-site scripting (XSS) and other code injection attacks.

explanation: It's like a set of rules that tells a web browser which scripts and resources are allowed to run on a website, preventing malicious code from being executed.

real-world examples: Defining a CSP header for a website that specifies allowed sources for scripts, stylesheets, and other content.

related terms: Web Application Security, Cross-Site Scripting (XSS), Web Browser Security

Continuous Adaptive Risk and Trust Assessment (CARTA)

definition: A security model continuously assessing and adjusting risk and trust levels in real-time.

explanation: It's like constantly checking and updating your security measures to keep up with new threats.

real-world examples: Adjusting authentication requirements based on user behavior or detecting and responding to new vulnerabilities as they emerge.

related terms: Risk Management, Adaptive Security, Zero Trust, Real-Time Monitoring

Control Objectives

definition: Specific aims that an organization's internal controls seek to achieve.

explanation: It's like goals set to ensure that processes run smoothly and securely.

real-world examples: Ensuring accurate financial reporting and safeguarding assets.

related terms: Internal Controls, Risk Management, Compliance

Control Plane

definition: The part of a network that manages how data is forwarded.

explanation: Think of it like an air traffic control tower, directing the flow of network traffic to ensure everything runs smoothly.

real-world examples: Routers and switches are essential components of the control plane.

related terms: Data Plane, Network Management, Routing Protocol

Control Self-assessment (CSA)

definition: A process through which an organization's controls are evaluated by the staff who are responsible for them.

explanation: It's like a self-check to ensure everything is in order before an official inspection.

real-world examples: Employees evaluating the effectiveness of internal financial controls.

related terms: Internal Audit, Risk Assessment, Compliance

Cookie

definition: A small text file stored on a user's computer by a website.

explanation: It's like a name tag that websites use to remember you when you visit.

real-world examples: Cookies are used to remember login information, shopping cart contents, and website preferences.

related terms: Web Browser, Tracking, Privacy

Cookie Hijacking

definition: An attack that steals a user's cookies to impersonate them online.

explanation: Imagine someone stealing your name tag at a party and using it to pretend to be you.

real-world examples: Attackers can use stolen cookies to access a victim's online accounts without their knowledge.

related terms: Session Hijacking, Man-in-the-Middle Attack, Cross-Site Scripting (XSS)

Cookie Poisoning

definition: An attack that modifies a cookie to inject malicious code or data.

explanation: It's like tampering with someone's name tag to include a hidden message or a tracking device.

real-world examples: Attackers can use poisoned cookies to track users' online activity or steal sensitive information.

related terms: Cross-Site Scripting (XSS), Web Application Security, Session Hijacking

Corporate Espionage

definition: The act of spying on a company to steal trade secrets, business plans, or other confidential information.

explanation: It's like a corporate James Bond, infiltrating a rival company to steal their secret formula.

real-world examples: Attackers may use a variety of tactics, including hacking, social engineering, and insider threats.

related terms: Cyber Espionage, Intellectual Property Theft, Trade Secret Theft

Credential Dumping

definition: The process of extracting usernames, passwords, and other credentials from a compromised system.

explanation: It's like stealing a master key ring that can unlock multiple doors.

real-world examples: Attackers use credential dumping tools to steal login credentials from memory, databases, or configuration files.

related terms: Pass-the-Hash, Golden Ticket Attack, Mimikatz

Credential Harvesting

definition: The process of collecting login credentials from users through various techniques, often for malicious purposes.

explanation: It's like a fisherman casting a wide net to catch as many fish as possible.

real-world examples: Phishing emails, keyloggers, and malware can all be used to harvest credentials.

related terms: Phishing, Malware, Keylogger

Credential Management

definition: The practice of securely storing, organizing, and managing user credentials.

explanation: It's like a password manager that keeps all your keys organized in one place, so you don't have to remember them all.

real-world examples: Using password managers, single sign-on (SSO), and other tools to manage user credentials securely.

related terms: Password Manager, Single Sign-On (SSO), Authentication

Credential Reuse

definition: The practice of using the same password across multiple sites, increasing the risk of compromise.

explanation: It's like using the same key for multiple doors.

real-world examples: Attackers gaining access to multiple accounts after a data breach.

related terms: Password Security, Identity Theft, Cybersecurity

Credential Stuffing

definition: An attack that uses stolen or hijacked credentials to try and gain access to multiple accounts.

explanation: It's like trying a stolen key on every door in a building, hoping to find one that it unlocks.

real-world examples: Attackers use lists of stolen usernames and passwords to try and gain access to online accounts.

related terms: Brute Force Attack, ATO, Password Reuse

Critical Security Control (CSC)

definition: A set of prioritized security measures designed to protect against the most common and dangerous cyberattacks.

explanation: It's like a list of the most important things to do to secure your home, like locking your doors, installing an alarm system, and keeping valuables out of sight.

real-world examples: Implementing firewalls, patching vulnerabilities, and using strong passwords.

related terms: Cybersecurity Framework, Security Best Practices, Risk Management

Cross-Site Request Forgery (CSRF)

definition: An attack that tricks a user into performing unwanted actions on a web application in which they are currently authenticated.

explanation: It's like a stranger using your phone to send a text message to someone without your knowledge or consent.

real-world examples: An attacker could send a malicious link to a victim that, when clicked, triggers an action on a website the victim is logged into, such as changing their email address or making a purchase.

related terms: Web Application Security, Session Hijacking, Injection Attack

Cross-Site Scripting (XSS)

definition: An attack that injects malicious code (usually JavaScript) into a website, which is then executed by other users' web browsers, potentially allowing the attacker to steal cookies, session tokens, or impersonate the user.

explanation: It's like a prankster hiding a whoopee cushion on a chair, waiting for someone to sit on it.

real-world examples: An attacker might inject a script into a website's comment section that steals other users' cookies or redirects them to a malicious website.

related terms: Web Application Security, Injection Attack, Vulnerability

Cross-Site Tracing (XST)

definition: A technique used to track users across different websites.

explanation: It's like a detective following a suspect's footprints across different locations.

real-world examples: Websites can use cookies, web beacons, or browser fingerprinting to track users' online activity.

related terms: Tracking, Privacy, Online Advertising

Cryptanalysis

definition: The study and practice of analyzing and breaking cryptographic systems.

explanation: It's like a detective solving a complex puzzle to uncover hidden messages.

real-world examples: Researchers breaking weak encryption algorithms to improve security standards.

related terms: Cryptography, Encryption, Decryption

Crypter

definition: A tool that obfuscates malware code to evade detection by security software.

explanation: It's like a disguise for malware, making it difficult for antivirus software to recognize and block it.

real-world examples: Cybercriminals use crypters to make their malware harder to detect.

related terms: Malware, Antivirus Evasion, Obfuscation

Crypto Locker

definition: A type of ransomware that encrypts a victim's files and demands a ransom payment in cryptocurrency to unlock them.

explanation: It's like a digital kidnapping of your files - the attackers hold them hostage until you pay up.

real-world examples: Crypto Locker was a notorious ransomware strain that emerged in 2013, encrypting files and demanding payment in Bitcoin.

related terms: Ransomware, Malware, Encryption

Crypto-Agility

definition: The capability of a system to switch between different cryptographic algorithms or protocols quickly and easily.

explanation: It's like a chameleon changing its colors to blend in with its surroundings. In this case, the chameleon is a cryptographic system adapting to new threats or vulnerabilities.

real-world examples: Used to mitigate the risk of a single cryptographic algorithm being compromised.

related terms: Cryptography, Encryption, Security

Cryptocurrency Security

definition: Measures and practices to protect cryptocurrency assets from theft and fraud.

explanation: It's like securing a digital wallet against pickpockets.

real-world examples: Using hardware wallets to store cryptocurrency securely.

related terms: Blockchain Security, Cybersecurity, Financial Security

Cryptographic Hash Function

definition: An algorithm that transforms any arbitrary block of data into a fixed-size string of characters, called a hash value.

explanation: It's like a fingerprint for data - it's unique to the original input and can be used to verify its integrity.

real-world examples: Used to verify the integrity of files, passwords, and digital signatures.

related terms: Hash Function, Checksum, Digital Signature

Cryptographic Key Management

definition: The process of managing cryptographic keys to ensure data security and integrity.

explanation: It's like keeping your keys in a secure place and making sure only the right people can use them.

real-world examples: Generating, storing, and distributing encryption keys for securing communications or data.

related terms: Encryption, Public Key Infrastructure (PKI), Key Exchange, Data Protection

Cryptographic Nonce

definition: An arbitrary number that is used only once in a cryptographic communication.

explanation: It's like a one-time password – it's generated randomly and never reused.

real-world examples: Used to prevent replay attacks, where an attacker intercepts and retransmits a message to gain unauthorized access.

related terms: Encryption, Cryptography, Replay Attack

Cryptographic Primitive

definition: A basic building block of a cryptographic algorithm.

explanation: It's like a brick in a wall – cryptographic primitives are combined to create more complex algorithms.

real-world examples: Hash functions, block ciphers, and stream ciphers.

related terms: Cryptography, Encryption, Algorithm

Cryptography

definition: The practice and study of techniques for securing communication in the presence of adversaries.

explanation: It's the art of secret writing – using codes and ciphers to protect messages from prying eyes.

real-world examples: Encryption, decryption, digital signatures, and authentication.

related terms: Cybersecurity, Information Security, Data Security

Cryptojacking

definition: The unauthorized use of someone else's computer to mine cryptocurrency.

explanation: It's like siphoning off your neighbor's electricity to power your own appliances.

real-world examples: Websites embedding malicious scripts that run in the background, using your computer's resources to mine cryptocurrency without your knowledge or consent.

related terms: Cryptocurrency, Malware, Browser-Based Mining

Crypto-Malware

definition: Malware that uses cryptography to encrypt files, hide its activities, or demand ransom payments.

explanation: It's like a criminal using a secret code to communicate and hide their activities from the authorities.

real-world examples: Ransomware, cryptojacking malware, and fileless malware that uses encryption to evade detection.

related terms: Malware, Encryption, Ransomware

Cryptomining

definition: The process of verifying and adding new transactions to a blockchain ledger in exchange for cryptocurrency rewards.

explanation: It's like digital gold mining - computers solve complex mathematical problems to "mine" new cryptocurrency coins.

real-world examples: Bitcoin mining, Ethereum mining.

related terms: Cryptocurrency, Blockchain, Proof of Work

Cryptomining Malware

definition: Malware that attackers secretly use in order to mine cryptocurrency from a victim's computer resources .

explanation: It's like someone secretly using your electricity to power their cryptocurrency mining operation.

real-world examples: Websites embedding malicious scripts that run in the background, consuming your computer's processing power to mine cryptocurrency.

related terms: Cryptojacking, Malware, Botnet

Cyber Attack

definition: A malicious attempt that damages, disrupts, or gains unauthorized access to a computer system or network.

explanation: It's like a digital break-in, where hackers try to steal information, disrupt operations, or cause damage to computer systems.

real-world examples: Hacking, phishing, malware infections, denial-of-service (DoS) attacks.

related terms: Cybersecurity, Threat, Vulnerability

Cyber Command

definition: A military organization responsible for planning, coordinating, and conducting cyber operations.

explanation: It's like a special forces unit for the digital battlefield, trained to defend against and launch cyberattacks.

real-world examples: The United States Cyber Command (USCYBERCOM) is responsible for defending U.S. military networks and conducting offensive cyber operations.

related terms: Cybersecurity, Cyber Warfare, National Security

Cyber Deception

definition: The use of deception techniques to mislead or confuse attackers.

explanation: It's like setting up a fake storefront to lure a thief away from the real valuables.

real-world examples: Decoy systems, honeypots, and misinformation campaigns.

related terms: Active Defense, Threat Intelligence, Cyber Threat Hunting

Cyber Espionage

definition: The practice of using cyber techniques to gather confidential information from governments, organizations, or individuals without permission.

explanation: It's like a spy secretly infiltrating and stealing sensitive information from an organization.

real-world examples: Nation-state actors hacking into government databases to steal classified information.

related terms: Hacking, Information Theft, Nation-state Attack

Cyber Hygiene

definition: Practices and steps that users and organizations take to maintain system health and to improve online security.

explanation: It's like regular cleaning and maintenance to keep your systems safe from threats.

real-world examples: Regularly updating software, using strong passwords, and backing up data.

related terms: Security Best Practices, System Maintenance, Threat Prevention, User Education

Cyber Incident

definition: An event that compromises the security of a computer system or network.

explanation: It's like a fire alarm going off in the digital world, signaling that something is wrong.

real-world examples: Data breaches, malware infections, system outages, or any other event that could harm an organization's information systems.

related terms: Cybersecurity, Incident Response, Data Breach

Cyber Insurance

definition: Insurance that helps organizations manage financial losses from cyberattacks and data breaches.

explanation: It's like insurance for your car, but instead of protecting against accidents, it protects against hackers and data thieves.

real-world examples: Covers costs associated with data breaches, ransomware attacks, and business interruption.

related terms: Cybersecurity, Risk Management, Insurance

Cyber Kill Chain

definition: A framework that describes the phases of a cyber attack, from reconnaissance to exfiltration.

explanation: It's like a playbook for hackers, outlining the steps they typically take to carry out an attack.

real-world examples: Understanding the kill chain can help defenders identify and stop attacks at earlier stages.

related terms: Threat Intelligence, Incident Response, Cyber Attack

Cyber Range

definition: A virtual environment used for cybersecurity training and testing.

explanation: It's like a flight simulator for pilots, but instead of practicing flying, you're practicing defending against cyberattacks.

real-world examples: Allows security professionals to simulate real-world attack scenarios in a safe and controlled environment.

related terms: Cybersecurity Training, Red Teaming, Ethical Hacking

Cyber Resilience

definition: The capability of an organization to prepare for, respond to, and recover from cyberattacks.

explanation: It's like having a strong immune system – it helps you fight off infections and recover quickly from illness.

real-world examples: Implementing a robust cybersecurity program, having a strong incident response plan, and regularly testing and updating your defenses.

related terms: Cybersecurity, Risk Management, Business Continuity

Cyber Resilience Framework

definition: A set of guidelines and best practices for building cyber resilience.

explanation: It's like a recipe for a healthy immune system, providing a roadmap for organizations to follow to improve their ability to withstand cyberattacks.

real-world examples: The NIST Cybersecurity Framework is a widely used example of a cyber resilience framework.

related terms: Cybersecurity, Risk Management, Business Continuity

Cyber Squad

definition: A team of cybersecurity professionals responsible for protecting an organization's systems and networks.

explanation: They are like the digital equivalent of a police force, patrolling the internet and investigating cybercrimes.

real-world examples: Conducting security assessments, responding to security incidents, and educating employees about cybersecurity best practices.

related terms: Cybersecurity, Incident Response, Security Operations Center (SOC)

Cyber Squatting

definition: The practice of registering, trafficking in, or using a domain name with bad faith intent to profit from the goodwill of a trademark belonging to someone else.

explanation: It's like buying a plot of land next to a popular tourist attraction and building a souvenir shop that uses the attraction's name.

real-world examples: Registering a domain name that is very similar to a popular brand's name, then trying to sell it back to them for a profit.

related terms: Domain Name, Trademark Infringement, Cybercrime

Cyber Threat Actor (CTA)

definition: An individual or group that conducts malicious activities in cyberspace.

explanation: They are the perpetrators of cyber attacks, ranging from script kiddies to sophisticated nation-state actors.

real-world examples: Hackers, hacktivists, cybercriminals, and state-sponsored actors.

related terms: Cybersecurity, Threat, Attacker

Cyber Threat Hunting

definition: The proactive search for cyber threats that are lurking within a network.

explanation: It's like having a security patrol actively looking for potential threats before they cause harm.

real-world examples: Security teams using advanced tools and techniques to identify and mitigate threats.

related terms: Threat Intelligence, Incident Response, Cybersecurity

Cyber Threat Information Sharing (CTIS)

definition: The practice of sharing information about cyber threats and vulnerabilities between organizations.

explanation: It's like a neighborhood watch group for the internet – everyone shares information about suspicious activity to help keep the entire community safe.

real-world examples: Organizations sharing threat intelligence with each other through platforms like Information Sharing and Analysis Centers (ISACs).

related terms: Threat Intelligence, Cybersecurity Collaboration, Information Sharing

Cyber Threat Intelligence (CTI)

definition: Information about current or emerging cyber threats, including tactics, techniques, and procedures (TTPs) used by attackers.

explanation: It's like a weather forecast for cyberattacks, providing information about potential threats so organizations can prepare and defend themselves.

real-world examples: Open-source intelligence (OSINT), threat feeds, and reports from security vendors and government agencies.

related terms: Threat Intelligence Platform, Cybersecurity, Risk Management

Cyber Vandalism

definition: The act of intentionally damaging, defacing, or disrupting digital property, such as websites, online services, or data, typically without financial gain.

explanation: It's like graffiti on a public building, where the goal is to cause visual damage or disruption rather than to steal or profit.

real-world examples: Website Defacement - Hackers changing the appearance of a website to display their own messages or images; or Service Disruption - Launching a denial-of-service (DoS) attack to make an online service temporarily unavailable.

related terms: Defacement, Denial of Service, Hacktivism

Cyberbullying

definition: The use of electronic communication, typically messages of an intimidating or threatening nature, to bully or harass a person.

explanation: It's like traditional bullying, but it happens online or through digital devices.

real-world examples: Spreading rumors online, sending mean text messages, or posting embarrassing photos of someone on social media.

related terms: Bullying, Harassment, Online Safety

Cybercrime

definition: Criminal activities carried out using computers or the internet.

explanation: It's any crime that involves a computer and a network.

real-world examples: Hacking, identity theft, online scams, cyberstalking, and the distribution of child pornography.

related terms: Cybersecurity, Computer Crime, Cyber Fraud

Cybercrime Investigation

definition: The process of investigating and gathering evidence of cybercrimes.

explanation: It's like detective work, but in the digital realm.

real-world examples: Tracking down hackers, analyzing malware, and recovering stolen data.

related terms: Digital Forensics, Cyber Law, Incident Response

Cybercrime-as-a-Service (CaaS)

definition: A business model where cybercriminals offer their services or tools to other criminals for a fee.

explanation: It's like a one-stop shop for cybercrime, where criminals can buy hacking tools, malware, or even hire hackers to carry out attacks.

real-world examples: Ransomware-as-a-Service (RaaS), phishing-as-a-service (PhaaS), and botnet rentals.

related terms: Cybercrime, Malware, Dark Web

Cyberdefense

definition: Measures taken to protect against cyberattacks.

explanation: It's like building and maintaining a fortress to keep out invaders.

real-world examples: Implementing firewalls and intrusion detection systems.

related terms: Cybersecurity, Risk Management, Threat Prevention

Cyberespionage

definition: The use of cyberattacks to steal classified or sensitive information from governments, organizations, or individuals.

explanation: It's like spying, but instead of using hidden cameras and microphones, attackers use computer networks and hacking tools.

real-world examples: Stealing government secrets, military plans, or intellectual property from businesses.

related terms: Cyber Warfare, Espionage, Nation-State Attack

Cybersecurity

definition: The practice of protecting systems, networks, and data from digital attacks.

explanation: It's like building a fortress around your digital assets to keep them safe from intruders.

real-world examples: Implementing firewalls, antivirus software, encryption, and strong passwords.

related terms: Information Security, Computer Security, Network Security

Cybersecurity Audit

definition: An evaluation of an organization's cybersecurity policies, procedures, and defenses.

explanation: It's like a security checkpoint ensuring all gates and barriers are functioning correctly.

real-world examples: Reviewing firewall configurations and access controls.

related terms: IT Audit, Compliance, Risk Management

Cybersecurity Awareness Month

definition: An annual campaign held every October to raise awareness about cybersecurity.

explanation: It's like a public service announcement for cybersecurity, reminding people to stay safe online.

real-world examples: Organizations host webinars, workshops, and other events to educate the public about cyber threats and how to protect themselves.

related terms: Cybersecurity Education, Security Awareness Training

Cybersecurity Capability Maturity Model (C2M2)

definition: A framework designed to help organizations evaluate and improve their cybersecurity capabilities.

explanation: It's like a roadmap for assessing and enhancing an organization's cybersecurity posture.

real-world examples: Organizations using C2M2 to identify strengths and weaknesses in their cybersecurity programs.

related terms: Cybersecurity Frameworks, Risk Management, Maturity Models

Cybersecurity Education

definition: The process of teaching individuals and organizations about cybersecurity risks and how to mitigate them.

explanation: It's like going to school to learn about cybersecurity - the curriculum covers everything from basic security hygiene to advanced threat detection and response.

real-world examples: Cybersecurity courses, certifications, and training programs.

related terms: Security Awareness Training, Cybersecurity Awareness

Cybersecurity Framework

definition: A set of standards, guidelines, and best practices in order to manage cybersecurity risk.

explanation: It's like a blueprint for building a secure house, outlining the materials, construction techniques, and safety features needed to keep the occupants safe.

real-world examples: The NIST Cybersecurity Framework, ISO 27001, and the CIS Controls.

related terms: Cybersecurity, Risk Management, Compliance

Cybersecurity Insurance

definition: Insurance that helps organizations manage financial losses from cyberattacks.

explanation: It's like car insurance, but instead of protecting against accidents, it protects against hackers and data breaches.

real-world examples: Covers costs associated with data breaches, ransomware attacks, and business interruption.

related terms: Cyber Insurance, Risk Management, Insurance

Cybersecurity Maturity Model

definition: A framework for determining the maturity of an organization's cybersecurity practices.

explanation: It's like measuring the readiness level of a security system.

real-world examples: Organizations using maturity models to benchmark and improve their cybersecurity practices.

related terms: Risk Management, Cybersecurity Frameworks, Compliance

Cybersecurity Maturity Model Certification (**CMMC**)

definition: A framework developed by the U.S. Department of Defense (DoD) to assess and certify the cybersecurity practices of defense contractors.

explanation: It's like a report card for cybersecurity, grading contractors on their ability to protect sensitive information.

real-world examples: Requires contractors to implement specific security controls and demonstrate their effectiveness.

related terms: Cybersecurity, Defense Industry, Compliance

Cybersecurity Mesh

definition: A flexible security architecture that protects different parts of a network independently.

explanation: It's like having multiple security fences around different areas to protect each one individually.

real-world examples: Using security measures that can move with devices and data across various environments, like in remote work setups.

related terms: Zero Trust, Distributed Security, Security Architecture, Network Security

Cybersecurity Operations

definition: The ongoing activities involved in protecting and defending an organization's systems and networks.

explanation: It's like a 24/7 security patrol, constantly monitoring for threats and responding to incidents.

real-world examples: Monitoring network traffic, detecting intrusions, analyzing logs, and patching vulnerabilities.

related terms: Security Operations Center (SOC), Incident Response, Threat Intelligence

Cybersecurity Posture

definition: The overall status of an organization's cybersecurity, including its defenses and readiness to respond to attacks.

explanation: It's like assessing the health and fitness of a security system.

real-world examples: Regularly assessing and updating security measures to maintain a strong cybersecurity posture.

related terms: Cybersecurity, Risk Management, Security Assessment

Cybersecurity Risk Management Framework (RMF)

definition: A structured approach to managing cybersecurity risk.

explanation: It's like a recipe for managing risk, with steps for identifying, assessing, and mitigating potential threats.

real-world examples: The NIST RMF is a popular framework used by many organizations.

related terms: Risk Management, Cybersecurity, Compliance

Cybersecurity Strategy and Implementation Plan (CSIP)

definition: A strategic plan outlining an organization's approach to managing cybersecurity risks.

explanation: It's like a game plan for building and maintaining a robust cybersecurity defense.

real-world examples: Developing a CSIP to guide cybersecurity investments and initiatives.

related terms: Cybersecurity, Risk Management, Strategic Planning

Cyberterrorism

definition: The use of cyberattacks to achieve political or social objectives by intimidating or coercing a government or civilian population.

explanation: It's like a terrorist attack, but instead of using bombs or guns, attackers use computers and networks.

real-world examples: Attacking critical infrastructure, disrupting financial markets, or spreading propaganda.

related terms: Terrorism, Cyber Warfare, Cybercrime

Cyberwarfare

definition: The use of cyberattacks by nation-states or their proxies to disrupt or damage the systems and networks of another country.

explanation: It's like traditional warfare but fought in the digital realm instead of on a physical battlefield.

real-world examples: Attacks on critical infrastructure, espionage, and disinformation campaigns.

related terms: Cyber Espionage, Nation-State Attack, Cybersecurity

Cypherpunk

definition: An individual who advocates for the use of strong cryptography and privacy-enhancing technologies as a means of social and political change.

explanation: They believe that privacy is a fundamental human right and that cryptography is a powerful tool to protect that right.

real-world examples: Developing and promoting open-source encryption tools, advocating for privacy-focused policies, and raising awareness about surveillance and censorship.

related terms: Cryptography, Privacy, Activism

DAD Triad

definition: A model used to describe the three primary goals of an attacker: Disclosure, Alteration, and Denial.

explanation: It's like the opposite of the CIA Triad (Confidentiality, Integrity, Availability), focusing on what attackers aim to achieve when they compromise a system.

real-world examples: Disclosure - An attacker stealing sensitive information (data breach), Alteration -An attacker changing data to cause harm or benefit themselves (data tampering), and Denial - An attacker preventing legitimate users from accessing a service (denial-of-service attack).

related terms: Cybersecurity, Threat Modeling, Attack Goals

Dark Web

definition: A part of the internet that is not indexed by search engines and can only be accessed using specialized software.

explanation: It's like a hidden part of the city where illegal activities take place.

real-world examples: Used for anonymous communication, the sale of illegal goods and services, and other illicit activities.

related terms: Deep Web, Tor, Onion Routing

Data Accountability and Trust Act (DATA)

definition: Proposed U.S. legislation aimed at establishing consumer data privacy rights and ensuring data security.

explanation: It's like a promise to consumers that their data will be handled with care and transparency.

real-world examples: Companies enhancing their data protection measures to comply with DATA requirements.

related terms: Data Privacy, Regulatory Compliance, Consumer Rights

Data Anonymization

definition: The method of removing personally identifiable information (PII) from data sets, so the individuals whom the data describe remain anonymous.

explanation: It's like blurring out faces in a photo to protect people's identities.

real-world examples: Anonymizing medical records for research purposes or masking customer data in analytics.

related terms: Data Privacy, Data Masking, Pseudonymization, GDPR

Data at Rest

definition: Data that is kept in storage media such as a hard drive or flash drive.

explanation: It's data that is not actively being transmitted or processed.

real-world examples: Files on your computer, data stored in a database, or information stored on a backup tape.

related terms: Data in Transit, Data in Use, Data Security

Data Backups

definition: Copies of data that are created for the purpose of restoring the original data in the event of loss or corruption.

explanation: It's like making copies of your important documents so you can recover them if the originals are lost or damaged.

real-world examples: Backing up your computer to an external hard drive or cloud storage service.

related terms: Data Recovery, Disaster Recovery, Backup Strategy

Data Breach

definition: An incident that results in the unauthorized access or disclosure of sensitive data.

explanation: It's like a thief breaking into your house and stealing your valuables.

real-world examples: A hacker gaining access to a company's customer database or a data leak caused by a misconfigured cloud server.

related terms: Data Loss, Cyber Attack, Cybersecurity Incident

Data Center Security

definition: The physical and virtual security measures implemented to protect data centers and their components.

explanation: Think of it like a high-security vault for your company's most valuable data assets.

real-world examples: Access controls, surveillance systems, fire suppression systems, and encryption of data at rest and in transit.

related terms: Physical Security, Network Security, Data Security

Data-Centric Security

definition: A security approach that focuses on protecting data itself rather than just the systems and networks that store and transmit it.

explanation: It's like putting a security wrapper around your data to protect it wherever it goes.

real-world examples: Encrypting data at rest and in transit and applying access controls directly to data files.

related terms: Data Encryption, Access Control, Data Loss Prevention (DLP), Information Security

Data Classification

definition: The process of categorizing data based on its sensitivity and value to an organization.

explanation: It's like sorting your belongings into different boxes based on their importance and value - some things you keep locked up, while others are easily accessible.

real-world examples: Classifying data as confidential, restricted, or public.

related terms: Data Security, Data Governance, Access Control

Data Classification Scheme

definition: A framework for classifying data based on its sensitivity and value.

explanation: It's like a set of guidelines for organizing your belongings into different boxes.

real-world examples: Government classification schemes (e.g., Top Secret, Secret, Confidential), industry-specific schemes (e.g., HIPAA for healthcare data), and custom schemes developed by organizations.

related terms: Data Classification, Data Security, Data Governance

Data Controller

definition: An entity that determines the purposes and means of processing personal data.

explanation: They are the ones who decide why and how personal data are collected, used, and stored.

real-world examples: A company that collects customer data for marketing purposes.

related terms: Data Protection, GDPR, Data Processor

Data Custodian

definition: An individual or team responsible for managing and protecting data on behalf of the data controller.

explanation: They are the caretakers of data, ensuring its security, integrity, and availability.

real-world examples: IT administrators, database administrators, and security teams.

related terms: Data Controller, Data Security, Data Governance

Data Destruction

definition: The process of permanently erasing data so that it cannot be recovered.

explanation: It's like shredding a document – once it's destroyed, it's gone for good.

real-world examples: Physically destroying hard drives, degaussing magnetic media, and using software to overwrite data.

related terms: Data Erasure, Data Sanitization, Data Disposal

Data Diddling

definition: The unauthorized modification of data before or during input into a computer system.

explanation: It's like a chef secretly changing the ingredients in a recipe to sabotage the dish.

real-world examples: Changing the amount of a transaction in a financial system or modifying the grades in a school's database.

related terms: Data Integrity, Data Manipulation, Fraud

Data Diode

definition: A network security device that allows one-way data flow, preventing data from flowing back in the opposite direction.

explanation: It's like a one-way valve that allows water to flow in one direction but not the other.

real-world examples: Used to protect sensitive networks from unauthorized access, such as industrial control systems or military networks.

related terms: Network Security, Air Gap, One-Way Transfer

Data Discovery

definition: The process of identifying and classifying sensitive data within an organization's systems and networks.

explanation: It's like taking inventory of your belongings to see what you have and where it's located.

real-world examples: Scanning databases, file servers, and cloud storage for sensitive data like personally identifiable information (PII) or financial data.

related terms: Data Classification, Data Loss Prevention (DLP), Data Security

Data Encryption

definition: The process of converting plaintext into ciphertext to protect its confidentiality.

explanation: It's like scrambling a message so that only authorized parties with the decryption key can read it.

real-world examples: Encrypting emails, files, and online transactions to prevent unauthorized access.

related terms: Encryption, Cryptography, Data Security

Data Encryption Standard (DES)

definition: A symmetric-key algorithm for encrypting data.

explanation: It was once a widely used standard but is now considered insecure due to its small key size.

real-world examples: DES has been replaced by more secure algorithms like AES.

related terms: Encryption, Cryptography, Symmetric Key Algorithm

Data Erasure

definition: The process of overwriting data on a storage device to make it unrecoverable.

explanation: It's like wiping a whiteboard clean – the old data is gone for good.

real-world examples: Using software tools to erase data from hard drives or other storage media before disposing of them.

related terms: Data Destruction, Data Sanitization, Data Wiping

Data Exfiltration

definition: The unauthorized transfer of data from a computer system or network.

explanation: It's like a thief stealing data from a bank vault.

real-world examples: Hackers stealing customer data from a company's servers or employees emailing confidential information to themselves.

related terms: Data Breach, Data Theft, Cyber Espionage

Data Governance

definition: A system of processes and controls in order to manage data availability, usability, integrity, and security.

explanation: It's like having a set of rules and procedures for managing a library, ensuring that books are organized, accessible, and in good condition.

real-world examples: Defining data policies, establishing data ownership, and implementing data quality controls.

related terms: Data Management, Data Security, Data Stewardship

Data Governance Act (DGA)

definition: A regulatory framework aimed at improving data governance and facilitating data sharing across the EU.

explanation: It's like a set of rules ensuring data is managed responsibly and can be shared securely.

real-world examples: Implementing data governance policies to enhance data quality and accessibility.

related terms: Data Management, Data Privacy, Regulatory Compliance

Data in Transit

definition: Data that is actively moving from one location to another, for example, over a network or through a communication channel.

explanation: It's like a package being shipped from one place to another.

real-world examples: Data being transmitted over the internet, between two computers, or through a wireless network.

related terms: Data at Rest, Data in Use, Encryption

Data Integrity

definition: The accuracy, completeness, and consistency of data over its entire lifecycle.

explanation: It's like making sure a recipe isn't altered so the final dish tastes as intended.

real-world examples: Implementing error detection and correction mechanisms, validating data input, and using checksums or hash functions to verify data integrity.

related terms: Data Security, Data Quality, Data Validation

Data Integrity Check

definition: A process for verifying that data has not been modified, corrupted, or tampered with.

explanation: It's like checking the seal on a package to ensure it hasn't been opened during transit.

real-world examples: Comparing a file's checksum or hash value with a known good value to verify its integrity.

related terms: Data Integrity, Checksum, Hash Function

Data Interception

definition: The act of capturing and monitoring data as it is transmitted over a network.

explanation: It's like eavesdropping on a conversation - you're listening in on data as it travels from one place to another.

real-world examples: Packet sniffing, man-in-the-middle (MitM) attacks.

related terms: Network Security, Eavesdropping, Man-in-the-Middle Attack

Data Leak

definition: The accidental or unintentional release of sensitive data to unauthorized parties.

explanation: It's like leaving your wallet on a park bench and someone finding it.

real-world examples: An employee accidentally emailing a confidential document to the wrong person or a company misconfiguring a cloud server, making data accessible to the public.

related terms: Data Breach, Data Loss, Human Error

Data Leakage

definition: The unauthorized transmission of data from within an organization to an external destination or recipient.

explanation: It's like a leaky faucet - sensitive data is slowly dripping out of the organization without anyone noticing.

real-world examples: Employees uploading confidential files to personal cloud storage accounts or hackers exfiltrating data through a hidden backdoor.

related terms: Data Exfiltration, Data Loss Prevention (DLP), Insider Threat

Data Loss Prevention (DLP)

definition: A strategy for ensuring that end users do not send sensitive or critical information outside the corporate network.

explanation: It's like a security guard at the company's data exit, checking outgoing information to ensure no sensitive details are leaving without permission.

real-world examples: Software that scans emails and attachments for sensitive data, blocks the transfer of certain file types, and monitors data movement in the cloud.

related terms: Data Leakage, Data Exfiltration, Endpoint Protection

Data Masking

definition: A technique that obscures specific data elements within a dataset to protect sensitive information.

explanation: It's like putting a mask over part of a document, so you can share it without revealing confidential details.

real-world examples: Replacing credit card numbers with "XXXX XXXX XXXX 1234" or email addresses with "[email address removed]" in test environments.

related terms: Data Anonymization, Data Obfuscation, Privacy

Data Minimization

definition: The principle of collecting and storing only the minimum amount of personal data needed for a specific purpose.

explanation: It's like packing light for a trip - only take what you absolutely need, leaving the unnecessary items at home.

real-world examples: Collecting only the customer data required to complete a transaction or deleting data that are no longer needed for business purposes.

related terms: Data Protection, Privacy by Design, GDPR

Data Normalization

definition: The process of organizing data in a database that reduces redundancy and improves data integrity.

explanation: It's like tidying up a messy closet - everything has its place, and there are no duplicates or inconsistencies.

real-world examples: Combining data from multiple sources into a single, consistent format, or removing duplicate entries from a database.

related terms: Database Management, Data Integrity, Data Quality

Data Privacy

definition: The right of individuals to control their personal information and how it is used.

explanation: It's like having a fence around your personal information, so only those you give to whom you give permission can see it.

real-world examples: controlling access to personal data, ensuring that it is used only for authorized purposes, and providing individuals with the ability to access and correct their data.

related terms: Data Protection, GDPR, Personally Identifiable Information (PII)

Data Privacy Framework

definition: A structured approach to managing data privacy risks and ensuring compliance with privacy regulations.

explanation: It's like a safety net protecting personal data from misuse and breaches.

real-world examples: Organizations adopting data privacy frameworks to comply with GDPR and CCPA.

related terms: Data Privacy, Risk Management, Compliance

Data Processing Agreement (DPA)

definition: A legal contract that outlines the obligations of a data controller and data processor in handling personal data.

explanation: It's like a prenuptial agreement for data, defining who is responsible for what in case of a data breach or other incident.

real-world examples: Required under GDPR when a company outsources data processing activities to a third-party vendor.

related terms: GDPR, Data Controller, Data Processor

Data Processor

definition: An entity that processes personal data on behalf of a data controller.

explanation: They are like the hired help who manage your data for you, but you (the data controller) are still ultimately responsible for it.

real-world examples: Cloud service providers, marketing agencies, and payroll processors are examples of data processors.

related terms: Data Controller, GDPR, Data Processing Agreement (DPA)

Data Profiling

definition: The process of analyzing data to understand its characteristics, content, and quality.

explanation: It's like taking a survey of your data to see what it looks like and how it's being used.

real-world examples: Identifying sensitive data, detecting anomalies, and assessing data quality.

related terms: Data Discovery, Data Governance, Data Management

Data Protection Impact Assessment (DPIA)

definition: A process to identify and mitigate privacy risks that are associated with data processing activities.

explanation: It's like a risk assessment for data privacy, evaluating potential harms and taking steps to minimize them.

real-world examples: Required under GDPR for certain high-risk processing activities, such as large-scale processing of sensitive data.

related terms: GDPR, Data Protection, Privacy Impact Assessment

Data Protection Officer (DPO)

definition: An individual appointed to oversee an organization's data protection strategy and compliance.

explanation: They are like the data protection police, making sure the company follows the rules and protects people's privacy.

real-world examples: Advising the company on data protection obligations, monitoring compliance, and acting as a contact point for data subjects.

related terms: GDPR, Data Protection, Privacy

Data Recovery

definition: The process of retrieving lost, deleted, or inaccessible data from storage media or other systems.

explanation: It's like salvaging valuables from a shipwreck.

real-world examples: Recovering data from a failed hard drive, accidentally deleted files, or a ransomware attack.

related terms: Data Loss, Data Backup, Disaster Recovery

Data Remanence

definition: The residual representation of data that remains on a storage device even after attempts have been made to remove or erase it.

explanation: It's like the faint traces of a message written on a chalkboard that can still be seen even after it's been erased.

real-world examples: Data can remain on hard drives, flash drives, and other storage media even after it's been deleted.

related terms: Data Destruction, Data Erasure, Data Sanitization

Data Retention

definition: The process of storing data for a specific period for legal, regulatory, or business purposes.

explanation: It's like keeping records of past transactions or activities for future reference.

real-world examples: Companies may need to retain customer data for a certain period to comply with regulations or for tax purposes.

related terms: Data Governance, Legal Hold, Data Retention Policy

Data Retention Policy

definition: A set of guidelines that defines how long an organization will keep different types of data.

explanation: It's like a schedule for cleaning out your closet, specifying which items you keep and for how long.

real-world examples: Defining how long customer data, financial records, and other types of information will be retained.

related terms: Data Retention, Data Governance, Records Management

Data Sanitization

definition: The process of removing or destroying data from a storage device to prevent it from being recovered.

explanation: It's like shredding a document - it makes the data unrecoverable even with specialized tools.

real-world examples: Using software to overwrite data on a hard drive or physically destroying the drive.

related terms: Data Destruction, Data Erasure, Data Wiping

Data Security

definition: The practice of protecting data from unauthorized access, corruption, or theft.

explanation: It's like building a fortress around your data to keep it safe from harm.

real-world examples: Encryption, access controls, firewalls, and intrusion detection systems.

related terms: Cybersecurity, Information Security, Data Protection

Data Sensitivity

definition: The level of risk associated with the unauthorized disclosure of certain data.

explanation: It's like classifying information as "top secret," "confidential," or "public" based on its potential impact if it falls into the wrong hands.

real-world examples: Personal data, financial data, and trade secrets are examples of sensitive data.

related terms: Data Classification, Data Protection, Risk Management

Data Sovereignty

definition: The principle that data is subject to the laws and regulations of the country in which it is stored.

explanation: It's like a country's borders extending into the digital realm - data stored within a country is subject to its laws.

real-world examples: GDPR, which applies to any company that processes the personal data of EU citizens, regardless of where the company is located.

related terms: Data Protection, Privacy, GDPR

Data Subject

definition: An individual who is the subject of personal data.

explanation: It's the person whose information is being collected, processed, or stored.

real-world examples: A customer, employee, or patient whose personal data are held by a company or organization.

related terms: Data Protection, GDPR, Personally Identifiable Information (PII)

Data Subject Access Request (DSAR)

definition: A request made by an individual to access their personal data that are held by an entity or organization.

explanation: It's like asking a company to show you the information they have about you.

real-world examples: Individuals have the right to request access to their data under GDPR and other data protection laws.

related terms: Data Protection, GDPR, Privacy Rights

Data Subject Rights

definition: The rights granted to individuals under data protection laws, such as the right to access, rectify, erase, and restrict processing of their personal data.

explanation: These rights empower individuals to control their personal information and ensure that it is handled responsibly by organizations.

real-world examples: Under GDPR, individuals can request to see their data, correct inaccuracies, have their data deleted, or object to its processing.

related terms: Data Protection, Privacy, GDPR

Data Tampering

definition: The malicious modification of data to compromise its integrity or availability.

explanation: Imagine someone secretly altering the ingredients in a recipe to ruin the dish.

real-world examples: Changing financial records, altering medical data, or manipulating scientific research.

related terms: Data Integrity, Cybercrime, Fraud

Database Activity Monitoring (DAM)

definition: A security technology that monitors and analyzes database activity to detect unauthorized access, suspicious behavior, or policy violations.

explanation: It's like having a security camera in your database, recording who accessed what and when.

real-world examples: Monitoring SQL queries, user logins, and data modifications to identify potential security threats.

related terms: Database Security, Intrusion Detection System (IDS), Security Information and Event Management (SIEM)

Database Firewall

definition: A security system designed to protect databases from unauthorized access and malicious activity.

explanation: It's like a security guard for your database, inspecting all incoming traffic and blocking anything suspicious.

real-world examples: Filtering SQL queries, blocking unauthorized access attempts, and preventing SQL injection attacks.

related terms: Database Security, Firewall, Intrusion Prevention System (IPS)

Database Security

definition: The practice of protecting and safeguarding databases from unauthorized access, use, disclosure, disruption, modification, or destruction.

explanation: It's like building a fortress around your database to keep out intruders and protect the valuable information within.

real-world examples: Implementing access controls, encryption, backups, and monitoring tools.

related terms: Cybersecurity, Data Security, Database Activity Monitoring (DAM)

Decentralized Identity

definition: A model for identity management where individuals own and control their digital identities without relying on a central authority.

explanation: It's like having your own personal ID that you control rather than relying on a government or company to manage it.

real-world examples: Using blockchain technology to create secure, self-sovereign identities that can be verified without a central authority.

related terms: Blockchain, Self-Sovereign Identity, Identity Management, Digital Identity

Deception Technology

definition: The use of decoys, traps, and misinformation to confuse and deter attackers.

explanation: It's like setting up a fake treasure chest to lure pirates away from the real treasure.

real-world examples: Honeypots, honeytokens, and fake vulnerabilities that trick attackers into revealing themselves.

related terms: Active Defense, Cyber Threat Hunting, Intrusion Detection System (IDS)

Decryption

definition: The process of converting ciphertext back into plaintext.

explanation: It's like translating a secret code back into plain language.

real-world examples: Decrypting an email message using your private key or unlocking a file with a password.

related terms: Encryption, Cryptography, Cipher

Decryption Oracle

definition: A system or service that can decrypt ciphertext, often used to test the security of encryption algorithms.

explanation: It's like a Rosetta Stone for encrypted messages, providing a way to decipher their meaning.

real-world examples: Researchers use decryption oracles to find weaknesses in encryption algorithms.

related terms: Cryptography, Encryption, Cryptanalysis

Deep Packet Inspection (DPI)

definition: A network security technique that examines the data payload of network packets to identify and classify traffic.

explanation: It's like a customs agent opening your suitcase to inspect its contents.

real-world examples: Used to filter out malicious traffic, enforce content policies, and prioritize network traffic based on its type or content.

related terms: Firewall, Network Security, Traffic Analysis

Deep Web

definition: The part of the internet that is not indexed by traditional search engines and including private databases and protected websites.

explanation: It's like the hidden part of an iceberg beneath the surface of the water.

real-world examples: Online banking portals and academic databases that require login credentials.

related terms: Dark Web, Internet, Privacy

Deepfake

definition: A video, audio, or image that has been manipulated using artificial intelligence to appear authentic but is actually fake.

explanation: It's like a photorealistic mask that someone wears to impersonate someone else.

real-world examples: Deepfakes can be used to spread misinformation, manipulate public opinion, or commit fraud.

related terms: Artificial Intelligence (AI), Machine Learning, Disinformation

Defacement

definition: An attack that alters the appearance of a website or webpage.

explanation: It's like graffiti artists vandalizing a public building with their own message.

real-world examples: Hackers replacing a website's homepage with their own message or propaganda.

related terms: Web Application Security, Cyber Vandalism, Hacktivism

Default Password

definition: A pre-set password that comes with a new device or software.

explanation: It's like the factory-installed lock on a new house - you should change it as soon as possible.

real-world examples: Default passwords are often easy to guess and pose a significant security risk.

related terms: Password Security, Weak Password, Credential Stuffing

Defense Advanced Research Projects Agency (DARPA)

definition: A U.S. government agency responsible for developing emerging technologies for military use.

explanation: It's like a research and development lab for the military, working on cutting-edge technologies that could be used on the battlefield.

real-world examples: DARPA has funded research that led to the development of the internet, GPS, and stealth technology.

related terms: Cybersecurity, Military Technology, Research and Development

Defense in Depth

definition: A security strategy that uses multiple layers of security controls to protect a system or network.

explanation: It's like a castle with multiple walls, moats, and guards - even if one layer of defense is breached, there are others to stop the attacker.

real-world examples: Combining firewalls, intrusion detection systems, antivirus software, and other security measures to create a layered defense.

related terms: Cybersecurity, Layered Security, Security Strategy

Degaussing

definition: The process of using a strong magnetic field to delete data from magnetic media.

explanation: It's like scrambling the magnetic particles on a hard drive or tape, making the data unreadable.

real-world examples: Used to securely erase hard drives, tapes, and other magnetic media before disposal.

related terms: Data Destruction, Data Erasure, Data Sanitization

Demilitarized Zone (DMZ) Host

definition: A server located in the DMZ that provides services to external users.

explanation: It's like a diplomat who is allowed to travel between two countries but is not fully trusted by either side.

real-world examples: Web servers, email servers, and DNS servers are often placed in the DMZ.

related terms: DMZ, Bastion Host, Perimeter Network

Denial of Service (DoS)

definition: An attack that floods a system or network with traffic, making it not available to legitimate users.

explanation: It's like blocking the entrance to a store with a crowd of people, preventing customers from getting inside.

real-world examples: Flooding a website with requests, sending a large number of emails to a server, or overloading a network with traffic.

related terms: Denial-of-Service Attack, Distributed Denial of Service (DDoS), Botnet

Denial of Service Attack

definition: A deliberate attempt to make a machine or network resource unavailable to its intended users, like by temporarily or indefinitely disrupting services of a host connected to the Internet.

explanation: An attack that aims to overwhelm a website or online service with traffic, causing it to crash or become unavailable to users.

real-world examples: SYN floods, ping floods, and other types of attacks that overwhelm a system's resources.

related terms: DoS, DDoS, Botnet

Denial-of-Service as a Service (DDoSaaS)

definition: A service that allows attackers to launch DDoS attacks for a fee.

explanation: Think of it like a cybercriminal "renting" a botnet to launch a DDoS attack.

real-world examples: These services often operate on the dark web and offer various levels of attack power and duration.

related terms: DDoS, Botnet, Cybercrime

DevSecOps

definition: An approach that incorporates security practices within the DevOps process.

explanation: It's like adding security checks at every stage of software development, from planning to deployment.

real-world examples: Automating security tests within continuous integration/continuous deployment (CI/CD) pipelines or ensuring code is reviewed for security vulnerabilities before deployment.

related terms: DevOps, CI/CD, Application Security, Automation

Dictionary Attack

definition: A method that involves trying a list of common words and phrases to crack a password.

explanation: It's like trying to guess someone's password by using a dictionary, starting with the most common words.

real-world examples: Attackers often use automated tools that can try thousands of passwords per second.

related terms: Brute Force Attack, Password Cracking, Wordlist

Dictionary Wordlist

definition: A method of password cracking that tries a pre-arranged list of common words and phrases.

explanation: Think of it like a burglar trying every word in the dictionary to guess your password.

real-world examples: Software that systematically tries common passwords like "password123" or "qwerty".

related terms: Brute Force Attack, Password Cracking, Credential Stuffing

Differential Cryptanalysis

definition: A method of breaking cryptographic algorithms by analyzing how small changes in the input affect the output.

explanation: Imagine a detective trying to decipher a code by looking for patterns in how the letters are substituted.

real-world examples: Used to break older encryption algorithms like DES.

related terms: Cryptanalysis, Encryption, Cipher

Digital Certificate

definition: An electronic document that verifies the identity of a website or server.

explanation: It's like a digital passport for a website, issued by a trusted authority (Certificate Authority) to confirm its authenticity.

real-world examples: When you see the padlock icon in your browser's address bar, it means the website is using a digital certificate.

related terms: SSL/TLS, Public Key Infrastructure (PKI), Certificate Authority (CA)

Digital Forensics

definition: The science of collecting and analyzing digital evidence to investigate cybercrimes.

explanation: It's like a CSI investigation but for computers.

real-world examples: Examining hard drives, network traffic logs, and other digital evidence to uncover traces of illegal activity.

related terms: Cybercrime Investigation, Incident Response, Evidence Collection

Digital Identity Verification

definition: The process of verifying that a person's digital identity matches their real-world identity.

explanation: It's like checking someone's ID to ensure they are who they say they are online.

real-world examples: Using facial recognition or biometric data to verify a user during account creation or login.

related terms: Identity Verification, Authentication, Biometrics, Multi-Factor Authentication (MFA)

Digital Millennium Copyright Act (DMCA)

definition: A U.S. law that criminalizes the production and dissemination of technology that can circumvent copyright protection.

explanation: It's like a law that prohibits picking locks – even if you're not using it to break into someone's house.

real-world examples: The DMCA is often used to take down websites that share copyrighted material without permission.

related terms: Copyright Law, Intellectual Property, Piracy

Digital Millennium Copyright Act (DMCA) Takedown Notice

definition: A legal notification sent to a website or service provider requesting the removal of copyrighted material.

explanation: It's like a cease-and-desist letter for copyright infringement.

real-world examples: A copyright holder might send a DMCA takedown notice to YouTube to remove a video that uses their music without permission.

related terms: DMCA, Copyright Infringement, Intellectual Property

Digital Operational Resilience Act (DORA)

definition: Proposed EU legislation aimed at ensuring financial entities can withstand, respond to, and recover from ICT-related disruptions.

explanation: It's like building a fortress to protect financial institutions from digital threats.

real-world examples: Financial institutions enhancing their resilience to cyberattacks and IT failures.

related terms: Cybersecurity, Risk Management, Regulatory Compliance

Digital Rights Management (DRM)

definition: Technologies that protect digital content from unauthorized copying and distribution.

explanation: It's like a lock on a digital book or movie, preventing you from sharing it with others.

real-world examples: DRM is used in software, music, movies, and e-books to prevent piracy.

related terms: Copyright Protection, Copy Protection, Digital Content

Digital Risk Protection (DRP)

definition: Services that protect against threats from the internet, including social media and the dark web.

explanation: It's like having a security system that watches the internet for any threats against your organization.

real-world examples: Detecting fake websites that try to steal company information or finding stolen credentials on the dark web.

related terms: Threat Intelligence, Dark Web Monitoring, Cyber Threat Intelligence (CTI), External Threat Protection

Digital Signature

definition: A mathematical technique used to validate the authenticity and integrity of a digital message or document.

explanation: It's like a handwritten signature, but for digital documents. It proves that the document was created by the person who claims to have created it and that it hasn't been tampered with.

real-world examples: Used to verify the authenticity of software updates, emails, and other digital documents.

related terms: Public Key Infrastructure (PKI), Cryptography, Authentication

Directive on Security of Network and Information Systems (NIS Directive)

definition: An EU directive targeted at improving the overall security of network and information systems across the Union.

explanation: It's like a set of rules ensuring critical infrastructure and services are protected against cyber threats.

real-world examples: Member states implementing measures to enhance the security of essential services.

related terms: Cybersecurity, Critical Infrastructure, Regulatory Compliance

Directory Harvest Attack

definition: An attack that attempts to gather valid email addresses from a domain.

explanation: It's like a telemarketer trying to build a list of potential customers by calling random phone numbers.

real-world examples: Attackers use automated tools to guess email addresses based on common naming conventions.

related terms: Spam, Phishing, Social Engineering

Directory Services

definition: A software system that stores, organizes, and provides access to information about a network's resources.

explanation: It's like a phonebook for a computer network, listing all the users, computers, printers, and other resources.

real-world examples: Active Directory, LDAP, and Novell eDirectory.

related terms: Network Management, Identity Management, Authentication

Directory Traversal Attack

definition: An attack that exploits a vulnerability in a web application to access files and directories outside of the intended web root directory.

explanation: It's like a burglar finding a secret passage in a house that leads to a hidden room.

real-world examples: Attackers can use directory traversal to access sensitive files, configuration files, or even execute commands on the server.

related terms: Web Application Security, Vulnerability, File Inclusion Vulnerability

Disaster Recovery (DR)

definition: The process of restoring an organization's IT infrastructure and data after a disaster.

explanation: It's like having a backup plan for your computer in case it crashes or is destroyed.

real-world examples: Restoring data from backups, setting up temporary offices, and rerouting network traffic.

related terms: Business Continuity Planning, Data Backup, High Availability

Disaster Recovery as a Service (DRaaS)

definition: A cloud-based service that provides disaster recovery capabilities.

explanation: It's like renting a backup office space in case your primary office is destroyed - everything you need is already there, ready to go.

real-world examples: Replicating data and applications to a cloud provider's infrastructure, so they can be quickly restored in the event of a disaster.

related terms: Disaster Recovery, Cloud Computing, Business Continuity Planning

Disaster Recovery Planning (DRP)

definition: The process of preparing for and recovering from disruptive events.

explanation: It's like having a contingency plan for when things go wrong.

real-world examples: Creating backup and recovery procedures for critical IT systems.

related terms: Business Continuity Planning (BCP), Risk Management, Incident Response

Disk Encryption

definition: The process of encrypting data on a hard drive or other storage device.

explanation: It's like putting a lock on your hard drive, so only authorized users with the decryption key can access the data.

real-world examples: BitLocker, FileVault, and VeraCrypt.

related terms: Encryption, Data Security, Full Disk Encryption (FDE)

Distributed Denial of Service (DDoS)

definition: An attack that floods a system or network with traffic from multiple sources, making it unavailable to legitimate users.

explanation: It's like a mob of people blocking the entrance to a store, preventing customers from getting inside.

real-world examples: Attackers often use botnets (networks of compromised computers) to launch DDoS attacks.

related terms: Denial of Service (DoS), Botnet, Network Security

Distributed Reflection Denial of Service (DRDoS)

definition: A type of DDoS attack where the attacker utilizes reflection techniques to amplify the attack traffic.

explanation: It's like using mirrors to blind someone with reflected light.

real-world examples: Attacks using DNS servers to amplify traffic and overwhelm targets.

related terms: DDoS, Cyberattack, Network Security

DMZ (Demilitarized Zone)

definition: A network segment that is located between an organization's internal network and the internet.

explanation: It's like a buffer zone between two warring countries, designed to prevent direct conflict.

real-world examples: Used to host publicly accessible servers, such as web servers and email servers, while keeping the internal network secure.

related terms: Network Security, Perimeter Network, Bastion Host

DNS Amplification Attack

definition: A type of DDoS attack that exploits vulnerabilities in the Domain Name System (DNS) to amplify the amount of traffic directed at a victim.

explanation: It's like using a megaphone to amplify your voice – the attacker sends a small request to a DNS server, but the server responds with a much larger amount of data, overwhelming the victim's network.

real-world examples: Attackers often use open DNS resolvers to launch DNS amplification attacks.

related terms: DDoS Attack, DNS, Botnet

DNS Blackhole

definition: A DNS server that gives a non-routable IP address in response to queries for certain domains.

explanation: It's like a black hole that absorbs any requests for those domains, preventing users from accessing them.

real-world examples: Used to block access to malicious websites or to prevent botnets from communicating with their command-and-control servers.

related terms: DNS, DNS Sinkhole, Botnet Mitigation

DNS Cache

definition: A temporary database that keeps information about recent DNS lookups.

explanation: It's like a memory bank for your computer or network, remembering the IP addresses of websites you've recently visited.

real-world examples: Speeds up web browsing by avoiding the need to query a DNS server every time you visit a website.

related terms: DNS, DNS Resolver, Domain Name System (DNS)

DNS Cache Poisoning

definition: An attack that redirects users to malicious websites by injecting false information into a DNS cache.

explanation: It's like tampering with a phone book so that when you look up a friend's number, you're actually given the number of a stranger.

real-world examples: Attackers can use DNS cache poisoning to redirect users to phishing websites or malware distribution sites.

related terms: DNS Spoofing, Pharming, Man-in-the-Middle Attack

DNS Filtering

definition: The practice of controlling which websites or domains can be accessed through the Domain Name System (DNS).

explanation: It's like a traffic cop directing internet traffic, allowing access to safe websites and blocking harmful or inappropriate ones.

real-world examples: Schools and businesses use DNS filtering to protect users from malicious websites and inappropriate content.

related terms: Web Filtering, Content Filtering, Parental Controls

DNS Flood

definition: A type of Denial of Service (DoS) attack that overwhelms a DNS server with a flood of DNS requests.

explanation: It's like a mob of people rushing into a restaurant all at once, overwhelming the staff and preventing legitimate customers from being served.

real-world examples: Attackers use botnets or other tools to send a massive number of DNS queries to a target server, causing it to crash or become unresponsive.

related terms: DoS Attack, DDoS Attack, Botnet

DNS Hijacking

definition: An attack that redirects a website's traffic to a fake website controlled by the attacker.

explanation: It's like a thief changing the road signs so you end up at the wrong destination.

real-world examples: Attackers can hijack DNS to steal login credentials, spread malware, or redirect users to malicious websites.

related terms: DNS Spoofing, Pharming, Man-in-the-Middle (MitM) Attack

DNS Poisoning

definition: An attack that corrupts the Domain Name System (DNS) data, causing it to return incorrect results.

explanation: It's like tampering with a phone book so that when you look up a friend's number, you're given the wrong one.

real-world examples: Attackers can use DNS poisoning to redirect users to malicious websites or intercept their traffic.

related terms: DNS Spoofing, Pharming, DNS Cache Poisoning

DNS Resolver

definition: A server that translates domain names (e.g., "www.example.com") into IP addresses (e.g., "192.0.2.1").

explanation: It's like a directory assistance service for the internet, helping your computer find the right website.

real-world examples: When you type a website address into your browser, your computer asks a DNS resolver to find the corresponding IP address.

related terms: DNS, Domain Name, IP Address

DNS Sinkhole

definition: A DNS server that gives a non-routable IP address in response to queries for certain domains.

explanation: It's like a dead-end road on a map - any traffic that tries to reach that domain will go nowhere.

real-world examples: Used to block access to malicious or unwanted websites or to redirect traffic to a honeypot.

related terms: DNS, DNS Blacklisting, Botnet Mitigation

DNS Spoofing

definition: An attack that forges DNS responses to redirect traffic to a malicious website.

explanation: It's like a fake signpost that points you in the wrong direction.

real-world examples: Attackers can use DNS spoofing to redirect users to phishing websites or malware distribution sites.

related terms: DNS Hijacking, DNS Poisoning, Pharming

DNS Tunneling

definition: A technique that uses DNS to encapsulate and transmit data covertly.

explanation: It's like hiding a message inside a seemingly harmless package.

real-world examples: Attackers can use DNS tunneling to exfiltrate data or bypass security controls.

related terms: Data Exfiltration, Covert Channel, Malware

DNS Zone Transfer

definition: A process for replicating DNS data across multiple DNS servers.

explanation: It's like making copies of a library catalog and distributing them to different branches.

real-world examples: Not only allows for redundancy and fault tolerance but also poses a security risk if not properly secured.

related terms: DNS, Zone File, DNS Server

DNSSEC (Domain Name System Security Extensions)

definition: A set of extensions to the DNS protocol that provide authentication and integrity protection for DNS data.

explanation: It's like adding a tamper-proof seal to a document to ensure it hasn't been altered.

real-world examples: DNSSEC helps prevent DNS spoofing and poisoning attacks.

related terms: DNS, Cryptography, Security

Document Exploitation (DOCEX)

definition: The process of extracting information from documents for intelligence or investigative purposes.

explanation: It's like a detective analyzing a suspect's diary to find clues about their activities.

real-world examples: Used by intelligence agencies, law enforcement, and private investigators to gather information.

related terms: Open-Source Intelligence (OSINT), Information Gathering, Cyber Espionage

Domain Generation Algorithm (DGA)

definition: An algorithm used by malware to dynamically generate domain names for command-and-control (C2) servers.

explanation: It's like a criminal constantly changing their phone number to avoid being tracked.

real-world examples: Makes it difficult for security professionals to block access to C2 servers, as the domain names are constantly changing.

related terms: Malware, Botnet, Command and Control

Domain Keys Identified Mail (DKIM)

definition: An email authentication method that allows the receiver to verify that an email was indeed sent and authorized by the owner of that domain.

explanation: It's like a digital signature for email, confirming that the message hasn't been tampered with and that it comes from the sender it claims to be from.

real-world examples: Used to combat email spoofing and phishing.

related terms: Email Security, Sender Policy Framework (SPF), DMARC

Domain Name Registrar

definition: A company that manages internet domain name reservations.

explanation: It's like a real estate agent for domain names, helping you buy, sell, or transfer ownership of a domain.

real-world examples: GoDaddy, Namecheap, and Network Solutions.

related terms: Domain Name, Top-Level Domain (TLD), Domain Name System (DNS)

Domain Name System (DNS)

definition: A hierarchical and decentralized naming system for computers, services, or other resources connected to the Internet or a private network.

explanation: It's like a phonebook for the internet, translating human-readable domain names into machine-readable IP addresses.

real-world examples: Translating "www.google.com" into an IP address like "142.250.184.142" so your browser can load the website.

related terms: Domain Name, DNS Resolver, DNS Server

Domain Name System (DNS) Security

definition: Measures to protect the DNS infrastructure from cyberattacks.

explanation: It's like securing the phone book of the internet.

real-world examples: Implementing DNSSEC to prevent DNS spoofing and cache poisoning.

related terms: Network Security, Cybersecurity, Internet Security

Domain Privacy Protection

definition: A service that hides the personal information of domain name owners from the public WHOIS database.

explanation: It's like putting a privacy screen on your mailbox to prevent strangers from seeing your address.

real-world examples: Protects domain owners from spam, identity theft, and other forms of harassment.

related terms: Domain Name, WHOIS, Privacy

Domain Reputation

definition: A measure of a domain's trustworthiness based on its history and associations with malicious activity.

explanation: It's like a credit score for a website – the higher the reputation, the more trustworthy it is considered to be.

real-world examples: Email filters and web browsers use domain reputation to block spam and malicious websites.

related terms: Email Security, Web Filtering, Blacklist

Domain Shadowing

definition: A technique used by attackers to create subdomains under a legitimate domain without the owner's knowledge.

explanation: It's like building a hidden room in someone else's house without their permission.

real-world examples: Used to host phishing websites, malware distribution sites, and other malicious content.

related terms: Domain Name, Phishing, Malware

Domain Squatting

definition: The practice of registering a domain name that is similar to a trademark or brand name, often with the intent to sell it to the rightful owner for a profit.

explanation: It's like buying up all the land around a popular tourist attraction and charging people to enter.

real-world examples: Registering a domain name like "gooogle.com" or "facebok.c om" and then trying to sell it to Google or Facebook.

related terms: Domain Name, Trademark Infringement, Cybercrime

Dormant Account

definition: A user account that has been inactive for a certain period.

explanation: Imagine a gym membership that hasn't been used in months. Dormant accounts pose a security risk as they may be vulnerable to unauthorized access or takeover.

real-world examples: A former employee's email account that is still active but hasn't been used in months.

related terms: Account Management, Inactive Account, Privilege Management

DOS Attack

definition: An attack that floods a system or network with traffic, making it unavailable to legitimate users.

explanation: It's like blocking the entrance to a store with a crowd of people, preventing customers from getting inside.

real-world examples: SYN flood, ping flood, and other types of attacks that overwhelm a system's resources.

related terms: Denial-of-Service (DoS), Distributed Denial of Service (DDoS), Botnet

Double Tagging

definition: An attack technique where an attacker inserts multiple HTML tags into a webpage to manipulate its structure or content.

explanation: Think of it like adding extra layers of clothing to a mannequin - it changes the way it looks and can be used to hide malicious content.

real-world examples: Used in cross-site scripting (XSS) attacks to bypass security filters or inject malicious code.

related terms: Cross-Site Scripting (XSS), Web Application Security, HTML Injection

DoublePulsar

definition: An exploit framework developed by the Equation Group (an alleged NSA-linked hacking group) that allows attackers to remotely execute code on vulnerable systems.

explanation: It's like a skeleton key that can unlock multiple doors, allowing attackers to gain access to a wide range of systems.

real-world examples: Used in the WannaCry and NotPetya ransomware attacks.

related terms: Exploit, Zero-Day Vulnerability, Remote Code Execution (RCE)

Downgrade Attack

definition: An attack that forces a system to downgrade its security protocols to a weaker, more vulnerable version.

explanation: It's like tricking a bank into using an outdated security system that is easier to break into.

real-world examples: Attackers might exploit a downgrade attack to intercept and decrypt secure communications.

related terms: SSL/TLS, Security Protocol, Man-in-the-Middle Attack

Downstream Liability

definition: The legal responsibility of an organization for security breaches or data leaks that occur further down the supply chain.

explanation: It's like a domino effect - if one company in a supply chain has a security breach, it can affect other companies that rely on their products or services.

real-world examples: A software vendor might be held liable for a data breach if a vulnerability in their software is exploited by hackers to attack their customers.

related terms: Supply Chain Security, Third-Party Risk, Data Breach

Doxing

definition: The act of publicly disclosing private personal information about an individual or organization.

explanation: It's like publishing someone's home address, phone number, and social security number online without their permission.

real-world examples: Often used for harassment, intimidation, or revenge.

related terms: Cyberbullying, Privacy, Online Harassment

Doxware

definition: A type of malware that threatens to publish a victim's sensitive data unless a ransom is paid.

explanation: It's like blackmail, but with digital information instead of physical evidence.

real-world examples: Threatening to release a victim's private photos, emails, or browsing history unless they pay up.

related terms: Ransomware, Malware, Extortion

Drive-by Attack

definition: An attack that uses a vulnerability in a website or application without any interaction from the user.

explanation: It's like a thief breaking into your house while you're asleep.

real-world examples: A user visits a compromised website, and their computer is automatically infected with malware.

related terms: Malware, Web Application Security, Vulnerability

Drive-by Download

definition: A type of drive-by attack where malicious software is automatically downloaded and installed on a user's computer without their knowledge or consent.

explanation: It's like a booby trap - you trigger it just by visiting a website, and suddenly you're infected with malware.

real-world examples: Often occurs when visiting compromised websites or clicking on malicious links.

related terms: Drive-by Attack, Malware, Exploit Kit

Drop Attack

definition: A type of Denial of Service (DoS) attack where an attacker causes a target system to crash or become unresponsive by sending a large number of partial packets.

explanation: It's like a swarm of bees stinging a victim repeatedly - the individual stings might not be that harmful, but the cumulative effect can be overwhelming.

real-world examples: Overloading a server with incomplete requests, causing it to drop legitimate connections.

related terms: DoS Attack, DDoS Attack, Network Security

Dropper

definition: A type of malware that installs or "drops" other malicious software onto a victim's system.

explanation: It's like a delivery truck for malware - it carries the malicious payload and deploys it on the target system.

real-world examples: Droppers often evade detection by antivirus software because they don't contain the actual malicious code themselves.

related terms: Malware, Trojan, Payload

Dual-Homed Host

definition: A computer or device that has two network interfaces, typically connected to two different networks.

explanation: It's like a house with two front doors - it can be accessed from two different streets (networks).

real-world examples: Used to create a DMZ, where a server can be accessed from the internet while still being isolated from the internal network.

related terms: DMZ, Network Security, Bastion Host

Due Diligence

definition: The investigation or audit of a potential investment or product to confirm all facts.

explanation: It's like checking under the hood before buying a used car.

real-world examples: Conducting financial and legal checks before a merger or acquisition.

related terms: Risk Assessment, Compliance, Internal Audit

Dumpster Diving

definition: The act of searching through trash or discarded items to find sensitive information.

explanation: It's like a scavenger hunt, but instead of looking for treasure, you're looking for discarded documents, passwords, or other sensitive information.

real-world examples: Dumpster diving can be used to gain access to buildings, steal identities, or commit fraud.

related terms: Social Engineering, Physical Security, Identity Theft

Dynamic Analysis

definition: The process of analyzing a program's behavior by executing it in a controlled environment.

explanation: It's like observing an animal in its natural habitat to understand its behavior.

real-world examples: Used to detect and analyze malware, identify vulnerabilities in software, and evaluate the effectiveness of security controls.

related terms: Malware Analysis, Vulnerability Assessment, Sandbox

Dynamic Application Security Testing (DAST)

definition: A method of testing web applications for vulnerabilities by simulating real-world attacks.

explanation: It's like a stress test for a bridge, where engineers simulate heavy traffic or earthquakes to see how the bridge holds up.

real-world examples: Tools like OWASP ZAP and Burp Suite are used to perform DAST.

related terms: Web Application Security, Penetration Testing, Vulnerability Scanning

Eavesdropping

definition: The act of secretly listening to a private conversation or communications.

explanation: It's like spying on someone's phone call or listening in on a private meeting.

real-world examples: Wiretapping, packet sniffing, and unauthorized access to emails or instant messages.

related terms: Privacy, Surveillance, Wiretapping

Egress Filtering

definition: The practice of controlling outgoing traffic from a network to prevent data exfiltration and other malicious activity.

explanation: It's like a security guard checking bags as people leave a store to make sure they haven't stolen anything.

real-world examples: Firewalls and intrusion prevention systems can be used to implement egress filtering.

related terms: Data Loss Prevention (DLP), Network Security, Firewall

Elliptic Curve Cryptography (ECC)

definition: A type of public-key cryptography that utilizes elliptic curves over finite fields to generate keys.

explanation: Think of it like a super-efficient lock and key system. The keys are smaller and faster to compute, yet just as secure as traditional methods, making it ideal for mobile devices and other resource-constrained environments.

real-world examples: Used in Bitcoin and other cryptocurrencies, secure messaging apps, and digital signatures.

related terms: Public Key Cryptography, RSA, Key Exchange

Email Security

definition: The practice of protecting email accounts and communications from unauthorized access, loss, or compromise.

explanation: It's like having a security system for your mailbox to prevent theft and tampering.

real-world examples: Spam filtering, antivirus scanning, encryption, and phishing awareness training.

related terms: Cybersecurity, Phishing, Malware

Embedded Malware

definition: Malicious code that is hidden within the firmware or software of a non-computer device.

explanation: Imagine a virus hiding inside your smart refrigerator, waiting to infect other devices on your network.

real-world examples: Malware embedded in routers, cameras, or industrial control systems.

related terms: Firmware, Internet of Things (IoT), Cyber-Physical Systems

Embedded System Security

definition: The process of securing embedded systems (computers embedded within other devices) from cyberattacks.

explanation: It's like installing a security system in your car to protect it from being hacked.

real-world examples: Ensuring the security of medical devices, industrial control systems, and smart home devices.

related terms: Cybersecurity, IoT Security, Cyber-Physical Systems

Emergency Management (Response)

definition: The coordination and management of resources and responsibilities for dealing with all aspects of emergencies, particularly preparedness, response, and recovery.

explanation: It's like having a plan and team ready to handle any emergencies that arise.

real-world examples: Developing and executing a disaster recovery plan after a major cybersecurity breach.

related terms: Incident Response, Disaster Recovery, Crisis Management

Encapsulation

definition: The process of enclosing one type of data within another type of data.

explanation: It's like putting a letter inside an envelope - the envelope protects the letter's contents while it's in transit.

real-world examples: Used in networking to encapsulate data packets within different protocol layers.

related terms: Network Protocols, OSI Model, TCP/IP

Encapsulation Security Payload (ESP)

definition: An IPSec protocol that maintains confidentiality, integrity, and authentication for data packets.

explanation: It's like wrapping a package in multiple layers of security tape, bubble wrap, and a tamper-proof seal.

real-world examples: Used in VPNs and other secure network communications to protect data in transit.

related terms: IPSec, Encryption, Authentication

Encrypted Communication

definition: The transmission of data that has been scrambled using an encryption algorithm.

explanation: It's like speaking in a secret language that only the intended recipient can understand.

real-world examples: Sending encrypted emails, using HTTPS for secure web browsing, and making secure phone calls.

related terms: Encryption, Cryptography, Confidentiality

Encrypted Virus

definition: A type of malware that uses encryption to hide itself from detection by antivirus software.

explanation: It's like a virus wearing a disguise to avoid being recognized by the immune system.

real-world examples: Polymorphic viruses and metamorphic viruses use encryption to constantly change their code.

related terms: Malware, Antivirus Evasion, Polymorphic Code

Encryption

definition: The process of converting information or data into a code, especially to prevent unauthorized access.

explanation: It's like scrambling a message so that only someone with the secret decoder ring can read it.

real-world examples: Used to protect sensitive data, such as passwords, financial information, and personal records.

related terms: Cryptography, Cipher, Decryption

Encryption Algorithm

definition: A set of mathematical instructions used to encrypt and decrypt data.

explanation: It's like a recipe for making a secret code - it specifies the steps to take to transform plaintext into ciphertext and vice versa.

real-world examples: Advanced Encryption Standard (AES), RSA, Twofish.

related terms: Encryption, Cryptography, Cipher

Encryption Key

definition: A piece of information used by an encryption algorithm to encrypt or decrypt data.

explanation: It's like the combination to a safe – without the key, you can't access the contents.

real-world examples: Encryption keys can be symmetric (same key for encryption and decryption) or asymmetric (different keys for encryption and decryption).

related terms: Encryption, Cryptography, Key Management

Endpoint

definition: Any device that connects to a network, such as computers, smartphones, and tablets.

explanation: It's like the individual locks on doors in a building.

real-world examples: Using endpoint protection software to secure laptops and mobile devices.

related terms: Endpoint Security, Cybersecurity, Network Security

Endpoint Detection and Response (EDR)

definition: A security solution that continuously monitors endpoints (devices like laptops and servers) to detect and respond to cyber threats.

explanation: It's like a security camera for your endpoints, recording everything that happens so you can spot suspicious activity and respond quickly.

real-world examples: Detecting malware infections, unauthorized access attempts, and other security incidents.

related terms: Endpoint Security, Threat Detection, Incident Response

Endpoint Protection Platform (EPP)

definition: A suite of security tools designed to protect endpoints from cyber threats.

explanation: It's like a security system for your computer, with multiple layers of protection to keep out intruders.

real-world examples: Antivirus, anti-malware, firewall, and intrusion prevention system (IPS).

related terms: Endpoint Security, Cybersecurity, Malware Protection

Endpoint Security

definition: The practice of securing endpoints (devices like laptops, desktops, and mobile devices) from cyberattacks.

explanation: It's like putting a lock on your front door to keep your home safe.

real-world examples: Installing antivirus software, keeping operating systems and applications up to date, and educating users about security best practices.

related terms: Cybersecurity, Endpoint Protection Platform (EPP), Mobile Device Management (MDM)

End-to-End Encryption (E2EE)

definition: A system of communication where messages can be read by the communicating users only.

explanation: It's like sending a secret message in a locked box that only the recipient can open.

real-world examples: Popular messaging apps like Signal and WhatsApp use E2EE to protect user privacy.

related terms: Encryption, Privacy, Confidentiality

Enterprise Risk Assessment

definition: A comprehensive evaluation of risks across an organization.

explanation: It's like a full-body scan to detect any potential issues or vulnerabilities.

real-world examples: Identifying and assessing risks in business operations, IT, and financial processes.

related terms: Risk Management, Compliance, Internal Audit

Enterprise Risk Management (ERM)

definition: A structured approach to managing risk across an organization.

explanation: It's like a strategic game plan to handle any curveballs that might come your way.

real-world examples: Implementing risk management frameworks like COSO ERM.

related terms: Risk Assessment, Compliance, Governance

Environmental, Social, and Governance (ESG)

definition: Criteria used to evaluate a company's operations concerning sustainability and ethical impact.

explanation: It's like a report card grading a company on its environmental and social responsibilities.

real-world examples: Companies publishing annual ESG reports to showcase their sustainability efforts.

related terms: Corporate Social Responsibility (CSR), Compliance, Governance

Ephemeral Port

definition: A temporary port number that is used for a single communication session.

explanation: It's like a temporary phone number that you use for one call and then discard.

real-world examples: Used in protocols like TCP and UDP to establish connections between clients and servers.

related terms: Port, TCP, UDP

Ethical Hacker

definition: A cybersecurity professional who uses their skills to test and improve the security of systems and networks.

explanation: They are like the "good guys" of the hacking world, using their knowledge to help organizations find and fix vulnerabilities before the bad guys can exploit them.

real-world examples: Performing penetration tests, conducting vulnerability assessments, and developing security recommendations.

related terms: Penetration Testing, Vulnerability Assessment, White Hat Hacker

Ethical Hacking

definition: The practice of using hacking techniques for legitimate purposes, such as testing the security of systems and networks.

explanation: It's like a locksmith testing the security of a lock by trying to pick it – the goal is to find weaknesses so they can be fixed.

real-world examples: Penetration testing, vulnerability assessments, and security audits.

related terms: Penetration Testing, Vulnerability Assessment, Cybersecurity

Ethics and Compliance

definition: Adherence to moral principles and regulatory requirements.

explanation: It's like following both the letter and the spirit of the law.

real-world examples: Implementing training programs on ethical behavior and regulatory compliance.

related terms: Code of Conduct, Compliance Program, Corporate Governance

EU Cybersecurity Act

definition: An EU regulation that strengthens the European Union Agency for Cybersecurity (ENISA) and establishes a framework for cybersecurity certification.

explanation: It's like setting high standards for cybersecurity across the EU to protect digital infrastructure.

real-world examples: Organizations obtaining cybersecurity certifications to demonstrate compliance with EU standards.

related terms: Cybersecurity, Regulatory Compliance, Certification

Event Correlation

definition: The process of analyzing security events from multiple sources to identify patterns, trends, and relationships.

explanation: It's like a detective connecting the dots to solve a crime.

real-world examples: SIEM systems use event correlation to identify security incidents and anomalies.

related terms: Security Information and Event Management (SIEM), Log Analysis, Threat Detection

Event Log

definition: A record of events that occur on a computer system or network.

explanation: It's like a diary for your computer, recording everything that happens, from software installations to security alerts.

real-world examples: Windows Event Log, Syslog, and application-specific logs.

related terms: Log Analysis, Security Information and Event Management (SIEM), Incident Response

Event Log Management

definition: The process of collecting, storing, analyzing, and archiving event logs from various systems and applications.

explanation: Think of it like a security guard's notebook, recording all the events that happen on a computer or network.

real-world examples: Log management software like Splunk, Graylog, or ELK Stack can be used to collect, store, and analyze event logs.

related terms: Security Information and Event Management (SIEM), Log Analysis, Threat Detection

Evil Maid Attack

definition: A type of attack where an attacker gains physical access to a device and installs malicious software to steal data.

explanation: It's like a maid with malicious intent tampering with a guest's laptop while cleaning their hotel room.

real-world examples: An attacker installing a keylogger on a laptop left unattended in a hotel room.

related terms: Physical Security, Data Theft, Malware

Evil Twin

definition: A rogue Wi-Fi access point that masquerades as a legitimate one to lure users into connecting.

explanation: Imagine a coffee shop imposter with free Wi-Fi, but it's a trap set by hackers to steal your data.

real-world examples: Attackers set up evil twins to intercept traffic, steal login credentials, or spread malware.

related terms: Wi-Fi Security, Rogue Access Point, Man-in-the-Middle Attack (MITM)

Executable and Linkable Format (ELF)

definition: A common standard file format for executables, object code, shared libraries, and core dumps.

explanation: It's like the blueprint for a program, containing instructions for the computer to execute.

real-world examples: Most Linux and Unix-based systems use ELF files.

related terms: Executable File, Object Code, Shared Library

Executable File

definition: A file that contains a program that can be run by a computer.

explanation: It's like a recipe for a dish - it contains instructions that the computer can follow to perform specific tasks.

real-world examples: EXE files on Windows, APP files on macOS, and ELF files on Linux.

related terms: Software, Application, Malware

Exploit

definition: A piece of software, a chunk of data, or a sequence of commands that takes advantage of a vulnerability to occur on computer software, hardware, or something electronic (usually computerized) by causing unintended or unanticipated behavior.

explanation: It's like a skeleton key that can unlock a door that wasn't meant to be opened.

real-world examples: A piece of code that takes advantage of a buffer overflow vulnerability to execute arbitrary code on a computer.

related terms: Vulnerability, Zero-Day, Patch

Exploit Chain

definition: A series of exploits used in conjunction to compromise a system.

explanation: It's like a chain reaction - one exploit triggers another, leading to a full system compromise.

real-world examples: An attacker might use a phishing email to deliver malware, which then exploits a vulnerability in the operating system to gain elevated privileges.

related terms: Exploit, Vulnerability, Attack Vector

Exploit Database

definition: A repository of publicly known exploits for various software vulnerabilities.

explanation: It's like a library of attack tools that hackers can use to exploit known weaknesses in software.

real-world examples: The Exploit Database, Metasploit, and other online resources provide information about exploits.

related terms: Vulnerability, Exploit, Ethical Hacking

Exploit Framework

definition: A software platform that provides tools and modules for developing and deploying exploits.

explanation: It's like a toolbox for hackers, containing everything they need to create and execute attacks.

real-world examples: Metasploit is a popular exploit framework used by both ethical hackers and malicious actors.

related terms: Exploit, Penetration Testing, Vulnerability Research

Exploit Kit

definition: A collection of exploits that are packaged together and sold or distributed online.

explanation: It's like a kit for building a bomb - it contains all the necessary components for an attacker to launch an attack.

real-world examples: Exploit kits are often used in drive-by download attacks, where a user's computer is infected simply by visiting a compromised website.

related terms: Malware, Drive-by Download, Exploit

Exploit Mitigation

definition: The process of reducing the risk or impact of an exploit.

explanation: It's like putting up security cameras and bars on your windows to deter burglars.

real-world examples: Patching vulnerabilities, configuring security settings, and using intrusion prevention systems.

related terms: Vulnerability Management, Patch Management, Risk Mitigation

Exploit Mitigation Kit

definition: A collection of tools and techniques for mitigating the risk of known exploits.

explanation: It's like a first aid kit for cybersecurity – it provides a quick and easy way to patch up vulnerabilities and protect your systems.

real-world examples: Microsoft's Enhanced Mitigation Experience Toolkit (EMET) is an example of an exploit mitigation kit.

related terms: Exploit, Vulnerability, Patch

Exploitable Vulnerability

definition: A weakness in a system or software that can be exploited by an attacker to gain unauthorized access or cause harm.

explanation: It's like a loose brick in a wall that an attacker can use to pull the whole thing down.

real-world examples: Unpatched software, misconfigurations, and weak passwords can all be exploitable vulnerabilities.

related terms: Vulnerability, Exploit, Threat

Exposure

definition: The potential loss or damage to an asset resulting from a threat.

explanation: It's like the amount of money you could lose if your house burns down.

real-world examples: Data breaches, system outages, and reputational damage are examples of exposure.

related terms: Risk, Vulnerability, Asset

Extended Detection and Response (XDR)

definition: A security approach that integrates multiple security tools for a comprehensive defense.

explanation: It's like having an all-in-one security system that watches over your entire network, not just individual parts.

real-world examples: Combining data from devices, networks, and cloud services to better detect and respond to threats.

related terms: Endpoint Detection and Response (EDR), Security Information and Event Management (SIEM), Threat Detection, Incident Response

Extensible Authentication Protocol (EAP)

definition: A framework for authentication protocols used in wireless networks and Point-to-Point Protocol (PPP) connections.

explanation: It's like a set of blueprints for building different types of locks and keys.

real-world examples: EAP-TLS, EAP-TTLS, PEAP, and EAP-FAST.

related terms: Authentication, Wireless Security, PPP

Extensible Markup Language (XML)

definition: A markup language that defines a set of rules for encoding documents in a format that is both human-readable and machine-readable.

explanation: It's like a language for organizing and structuring data in a way that both humans and computers can understand.

real-world examples: Used in a wide range of applications, including web development, document management, and data exchange.

related terms: Markup Language, HTML, JSON

External Audit

definition: An independent evaluation of an organization's financial statements and practices.

explanation: It's like having an impartial referee review the game to ensure fair play.

real-world examples: Financial audits conducted by external accounting firms.

related terms: Internal Audit, Compliance, Financial Reporting

External Network Penetration Test

definition: A security assessment of an organization's network perimeter from an external perspective.

explanation: It's like a burglar trying to break into a house to test its security.

real-world examples: Ethical hackers simulate real-world attacks to identify vulnerabilities in the network perimeter.

related terms: Penetration Testing, Vulnerability Assessment, Ethical Hacking

External Vulnerability Scan

definition: A security assessment that scans an organization's external-facing systems and networks for vulnerabilities.

explanation: It's like a health checkup for your network perimeter, identifying any weaknesses that could be exploited by attackers.

real-world examples: Using automated tools to scan for open ports, outdated software, and misconfigurations.

related terms: Vulnerability Assessment, Penetration Testing, Security Scanning

Failed Login Attempt

definition: An unsuccessful attempt to log into a system or account.

explanation: It's like trying to open a door with the wrong key.

real-world examples: Failed login attempts can be caused by mistyped passwords, brute-force attacks, or other malicious activity.

related terms: Authentication, Brute Force Attack, Account Lockout

Fake Antivirus

definition: Malicious software that pretends to be legitimate antivirus software to trick users into installing it.

explanation: It's like a fake security guard letting thieves into a building.

real-world examples: Scareware that prompts users to buy unnecessary or harmful software.

related terms: Malware, Social Engineering, Cybercrime

False Alarm

definition: A security alert that is triggered by a harmless event or activity.

explanation: It's like a smoke detector going off when you burn toast.

real-world examples: Antivirus software flagging a legitimate file as malware, or an intrusion detection system alerting on normal network traffic.

related terms: False Positive, True Positive, Security Alert

False Negative

definition: A security event or threat that is not detected by a security system.

explanation: It's like a burglar alarm that fails to go off when someone breaks into your house.

real-world examples: Malware that evades detection by antivirus software, or an attacker who successfully infiltrates a network without triggering any alerts.

related terms: False Positive, True Negative, Security Incident

False Positive

definition: A test result that incorrectly indicates the presence of a condition or threat.

explanation: Imagine a smoke detector going off when you're just cooking dinner - it's annoying, but better than a missed fire. In cybersecurity, a false positive is a security alert that is triggered by harmless activity.

real-world examples: An antivirus software incorrectly flagging a legitimate program as malware, or a spam filter blocking a harmless email.

related terms: False Negative, True Positive, Security Alert

Fault Injection Attack

definition: An attack that introduces faults into a system to cause it to malfunction or reveal sensitive information.

explanation: It's like tampering with a car's engine to make it break down or reveal its inner workings.

real-world examples: Attackers might use voltage glitches or laser beams to manipulate the behavior of a chip or device.

related terms: Hardware Hacking, Side-Channel Attack, Glitch Attack

Federated Identity

definition: An arrangement that allows users to use the same identity across multiple systems or organizations.

explanation: It's like using your driver's license as identification for multiple services, such as renting a car or checking into a hotel.

real-world examples: Single sign-on (SSO) solutions, where you use one set of credentials to access multiple applications.

related terms: Single Sign-On (SSO), Identity Management, Authentication

Federated Identity Management (FIM)

definition: A system that manages federated identities across different domains or organizations.

explanation: It's like a central authority that verifies and manages the "passports" (digital identities) of people who travel between different countries (organizations).

real-world examples: Enabling users to access resources across multiple organizations with a single login.

related terms: Federated Identity, Identity Management, SSO

File Inclusion Vulnerability

definition: A vulnerability that lets an attacker include a file on a server that they shouldn't have access to.

explanation: It's like a burglar finding a hidden key that unlocks a secret room in a house.

real-world examples: Local File Inclusion (LFI) and Remote File Inclusion (RFI) attacks.

related terms: Web Application Security, Injection Attack, Vulnerability

File Integrity Monitoring (FIM)

definition: A security technology that monitors and detects unauthorized changes to files and directories.

explanation: It's like a security camera for your files, alerting you if someone tampers with them.

real-world examples: Used to detect malware infections, unauthorized changes to system files, or data breaches.

related terms: Intrusion Detection System (IDS), Security Information and Event Management (SIEM), Tripwire

File Transfer Protocol (FTP)

definition: A standard network protocol used to transfer files between a client and server on a computer network.

explanation: It's like a delivery service for files, allowing you to send and receive files over the internet.

real-world examples: Uploading files to a web server, downloading files from a remote computer, and sharing files between colleagues.

related terms: Network Protocol, File Sharing, SFTP

File Transfer Protocol Secure (FTPS)

definition: A secure version of the File Transfer Protocol (FTP) that uses SSL/TLS to encrypt data transfers.

explanation: It's like sending a file through a secure tunnel - the contents are protected from eavesdropping and tampering.

real-world examples: Used for transferring sensitive files, such as financial data or personal information.

related terms: FTP, Encryption, SSL/TLS

File Vault

definition: A disk encryption program included with macOS.

explanation: It's like a built-in safe for your Mac's hard drive, encrypting all the data so that only authorized users can access it.

real-world examples: Protects data on your Mac from unauthorized access if your device is lost or stolen.

related terms: Disk Encryption, Full Disk Encryption (FDE), Data Security

Fileless Attack

definition: A type of cyberattack that doesn't rely on conventional malware files.

explanation: It's like a ghost that haunts your computer, leaving no trace behind.

real-world examples: Attackers use scripts, legitimate tools, and memory-resident techniques to evade detection.

related terms: Malware, Advanced Persistent Threat (APT), Endpoint Detection and Response (EDR)

Fileless Malware

definition: Malicious software that operates in memory and doesn't write files to disk.

explanation: It's like a ninja that leaves no footprints - it's difficult to detect and remove because it doesn't exist as a physical file.

real-world examples: Often uses PowerShell, WMI, or other legitimate tools to execute commands and evade detection.

related terms: Malware, Fileless Attack, Advanced Persistent Threat (APT)

Fingerprinting

definition: The process of identifying a device or user based on unique characteristics or behaviors.

explanation: It's like a detective collecting fingerprints or footprints at a crime scene to identify the perpetrator.

real-world examples: Websites can use browser fingerprinting to track users, and network administrators can use device fingerprinting to identify unauthorized devices.

related terms: Device Fingerprinting, Browser Fingerprinting, Tracking

Firewalking

definition: A network security technique used to determine which ports on a firewall are open.

explanation: It's like knocking on different doors to see which ones are unlocked.

real-world examples: Used to test the effectiveness of a firewall and identify potential vulnerabilities.

related terms: Port Scanning, Network Security, Firewall

Firewall

definition: A network security system that monitors and controls incoming and outgoing network traffic based on predetermined security rules.

explanation: It's like a security guard for your network, checking every visitor's ID and only allowing those who are authorized to enter.

real-world examples: Used to protect networks from unauthorized access, malware, and other threats.

related terms: Network Security, Intrusion Prevention System (IPS), Security

Firewall Log

definition: A record of events and activities that occur on a firewall.

explanation: It's like a security guard's logbook, documenting who entered and exited the building and what they did while they were there.

real-world examples: Firewall logs can be used to detect security incidents, troubleshoot network problems, and monitor network traffic.

related terms: Log Analysis, Security Information and Event Management (SIEM), Firewall

Firewall Policy

definition: A set of rules that define how a firewall should handle different types of network traffic.

explanation: It's like a set of instructions for a security guard, telling them who to let in and who to keep out.

real-world examples: Firewall policies typically specify which ports are open, which protocols are allowed, and which IP addresses are trusted.

related terms: Firewall, Network Security, Access Control

Firewall Rule

definition: A single instruction within a firewall policy that specifies how a particular type of traffic should be handled.

explanation: It's like a single line in a security guard's instructions, telling them to allow traffic from a specific IP address or block traffic on a specific port.

real-world examples: A firewall rule might allow incoming HTTP traffic on port 80 but block all other traffic.

related terms: Firewall, Firewall Policy, Network Security

Firmware

definition: Permanent software programmed into a read-only memory.

explanation: It's like the operating system of a simple electronic device, providing the basic instructions for how it functions.

real-world examples: Firmware is found in devices like routers, printers, and cameras.

related terms: Software, Embedded System, Hardware

Firmware Analysis

definition: The process of examining the firmware of a device to identify vulnerabilities or malicious code.

explanation: It's like dissecting a frog to understand its anatomy. In this case, the "frog" is the firmware, and the goal is to find any hidden security flaws.

real-world examples: Used to identify vulnerabilities in IoT devices, routers, and other embedded systems.

related terms: Vulnerability Assessment, Reverse Engineering, Embedded System Security

Firmware Attacks

definition: Exploiting vulnerabilities in the firmware of devices in order to gain unauthorized access.

explanation: It's like tampering with the engine control module of a car to gain control.

real-world examples: Attacks on router firmware to intercept network traffic.

related terms: Hardware Security, Cybersecurity, Exploits

First-Party Cookie

definition: A cookie set by the website you're currently visiting.

explanation: It's like a name tag given to you by the host of a party.

real-world examples: Used to remember your login information, shopping cart contents, and other website preferences.

related terms: Cookie, Third-Party Cookie, Web Tracking

Flash Cookie

definition: A type of cookie that Adobe Flash Player stores on your computer.

explanation: It's like a hidden name tag that websites can use to track you across different websites, even if you delete your regular cookies.

real-world examples: Used to track user behavior, deliver targeted advertising, and store persistent data.

related terms: Cookie, Web Tracking, Privacy

Footprinting

definition: The process of gathering information about a target system in order to identify vulnerabilities.

explanation: It's like scouting the layout of a building before planning a break-in.

real-world examples: Using tools to gather data on a target's network structure, IP addresses, and available services.

related terms: Reconnaissance, Information Gathering, Penetration Testing

Forensic Analysis

definition: The process of using scientific knowledge for collecting, examining, and evaluating digital evidence to present facts in a court of law.

explanation: Think of it like a detective investigating a crime scene but in the digital world. They examine clues to piece together what happened and who was responsible.

real-world examples: Analyzing a hard drive for deleted files, examining network traffic logs, or recovering data from a damaged device.

related terms: Digital Forensics, Cybercrime Investigation, Incident Response

Forensic Image

definition: An exact bit-by-bit copy of a digital storage device, such as a hard drive.

explanation: It's like a photograph of a crime scene - it captures all the data at a specific point in time, preserving it for later analysis.

real-world examples: Used in digital forensics investigations to analyze evidence without altering the original data.

related terms: Digital Forensics, Evidence Collection, Data Acquisition

Forensic Investigator

definition: A specialist who examines and analyzes digital evidence to support legal investigations.

explanation: They are the detectives of the digital world, using specialized tools and techniques to uncover evidence of cybercrime.

real-world examples: Police detectives, cybersecurity experts, and private investigators who specialize in digital forensics.

related terms: Digital Forensics, Cybercrime Investigation, Incident Response

Formjacking

definition: Injecting malicious code into websites to steal user data entered into forms.

explanation: It's like a hidden skimmer on an ATM that captures card information.

real-world examples: Attackers injecting code into e-commerce websites to steal payment information.

related terms: Cybercrime, Web Application Security, Data Theft

Forward Secrecy

definition: A cryptographic property that ensures that even if a long-term key is compromised, past communication sessions remain secure.

explanation: Imagine each conversation you have using a unique, disposable key that is destroyed after the conversation ends. Even if someone steals your key ring later, they can't decrypt your past conversations.

real-world examples: Used in secure communication protocols like TLS to protect against future decryption of intercepted data.

related terms: Encryption, Cryptography, Perfect Forward Secrecy (PFS)

Forwarded Events

definition: Security events or alerts that are sent from one system to another for analysis or action.

explanation: It's like a neighborhood watch group where members share information about suspicious activity with each other.

real-world examples: A firewall might forward logs of suspicious traffic to a Security Information and Event Management (SIEM) system for further analysis.

related terms: Security Information and Event Management (SIEM), Log Management, Event Correlation

Fraud Detection

definition: The process of identifying and preventing fraudulent activities.

explanation: It's like a security camera system that catches thieves in the act.

real-world examples: Using algorithms to detect unusual credit card transactions.

related terms: Risk Management, Compliance, Internal Controls

Fraud Risk Management

definition: Strategies and processes to identify, assess, and mitigate the risk of fraud.

explanation: It's like setting traps and alarms to catch and prevent fraudsters.

real-world examples: Implementing controls to prevent and detect fraudulent activities in financial processes.

related terms: Fraud Detection, Internal Controls, Compliance

Full Disk Encryption (FDE)

definition: The process of encrypting all data on a hard drive or other storage device.

explanation: It's like putting a lock on your entire hard drive, so that even if someone steals it, they can't access the data without the correct password or key.

real-world examples: BitLocker (Windows), FileVault (macOS), and VeraCrypt are examples of FDE software.

related terms: Encryption, Data Security, Disk Encryption

Fuzz Testing / Fuzzing

definition: A software testing technique that involves providing invalid, unexpected, or random data as inputs to a computer program to test its robustness and identify potential vulnerabilities.

explanation: It's like throwing everything but the kitchen sink at a program to see if it breaks.

real-world examples: Fuzzing is commonly used to find bugs and security flaws in software.

related terms: Software Testing, Vulnerability Assessment, Security Testing

Gateway

definition: A network node that acts as an entrance to another network.

explanation: Think of it like a doorway between two rooms - it allows traffic to flow between different networks.

real-world examples: Routers and firewalls often act as gateways, connecting a local network to the internet.

related terms: Router, Firewall, Network Security

Gateway Firewall

definition: A firewall placed at the edge of a network, acting as a wall between the internal network and the internet.

explanation: It's like a guard at the gate of a castle, checking everyone who enters and leaves to ensure they are authorized.

real-world examples: Gateway firewalls filter incoming and outgoing traffic to protect the internal network from attacks.

related terms: Firewall, Network Security, Perimeter Security

General Data Protection Regulation (GDPR)

definition: A European Union law regulating and controlling personal data collection, use, and storage.

explanation: It's like a set of rules for how companies can handle your personal information, giving you more control over your data.

real-world examples: Requires companies to obtain consent before collecting personal data, provide individuals with access to their data, and implement appropriate security measures to protect data.

related terms: Data Protection, Privacy, European Union

Generic Routing Encapsulation (GRE)

definition: A tunneling protocol that can encapsulate various network layer protocols inside virtual point-to-point links.

explanation: It's like putting a smaller package inside a larger one for shipping.

real-world examples: Used to create virtual private networks (VPNs) and to route traffic over complex network topologies.

related terms: VPN, Tunneling Protocol, Network Routing

Geographic Information System (GIS)

definition: A system that captures, stores, manipulates, analyzes, manages, and presents spatial or geographic data.

explanation: It's like a digital map that can show you where things are located and how they relate to each other.

real-world examples: Used in urban planning, environmental management, and emergency response.

related terms: Mapping, Geospatial Data, Location Intelligence

Geolocation Spoofing

definition: The act of falsifying or manipulating location data to deceive others about your real location.

explanation: It's like changing your home address to a fake one on your GPS device to make it appear that you are in a different location. This technique can be used to bypass geographic restrictions on services or to mask your real location.

real-world examples: Can be used to bypass geographic restrictions, mask your identity, or even commit fraud.

related terms: Location Spoofing, GPS Spoofing, Privacy

Ghostware

definition: Malicious software designed to hide its presence and activities from detection tools.

explanation: It's like a ghost that moves through walls without being seen.

real-world examples: Advanced persistent threats (APTs) using ghostware to remain undetected in a network for long periods.

related terms: Stealth Malware, APT, Cybersecurity

Global Positioning System (GPS) Spoofing

definition: A type of geolocation spoofing that involves sending false GPS signals to a receiver, causing it to calculate an incorrect location.

explanation: It's like transmitting false GPS signals to a GPS receiver, causing it to report an incorrect location. This can disrupt navigation systems, mislead tracking services, and potentially cause significant confusion or harm.

real-world examples: Can be used to disrupt navigation systems, track individuals, or even cause accidents.

related terms: Geolocation Spoofing, GPS Jamming, Cyber Attack

Golden Ticket Attack

definition: A type of attack that exploits a vulnerability in the Kerberos authentication protocol to gain unauthorized access to a network.

explanation: It's like having a master key that unlocks every door in a building. In this context, attackers exploit a vulnerability in the Kerberos authentication protocol to create a fake ticket-granting ticket (TGT), allowing them unauthorized access to any resource within the network.

real-world examples: Attackers can use a stolen Kerberos ticket-granting ticket (TGT) to impersonate any user on the network.

related terms: Kerberos, Authentication, Privilege Escalation

Governance Framework

definition: A structure of rules and practices to ensure accountability, fairness, and transparency in an organization.

explanation: It's like the constitution that outlines how an organization is governed.

real-world examples: Corporate governance policies that define the roles and responsibilities of the board and management.

related terms: Corporate Governance, Compliance, Risk Management

Governance, Risk, and Compliance (GRC)

definition: A framework that manages an organization's overall governance, risk management, and regulatory compliance.

explanation: It's like a set of rules and procedures to ensure that a company is operating ethically, managing risks effectively, and complying with all applicable laws and regulations.

real-world examples: Implementing policies, procedures, and controls to manage risk, comply with regulations, and achieve organizational objectives.

related terms: Cybersecurity, Risk Management, Compliance

Gramm-Leach-Bliley Act (GLBA)

definition: A U.S. federal law requiring financial institutions to explain how they share and protect customer information.

explanation: It's like a transparency pledge ensuring customers know how their data is handled.

real-world examples: Banks implementing privacy notices and security measures to protect customer data.

related terms: Data Privacy, Financial Compliance, Consumer Protection

Gray Hat Hacker

definition: A hacker who may violate ethical standards or principles but does not have malicious intent.

explanation: A hacker who operates in a gray area between ethical and unethical hacking. They may violate laws or ethical standards but typically do not have malicious intent. However, their actions can still lead to legal consequences or unintended harm.

real-world examples: Might discover a vulnerability and publicly disclose it without notifying the affected vendor or hack into a system to prove a point.

related terms: Hacker, White Hat Hacker, Black Hat Hacker

Grayware

definition: Software potentially unwanted or harmful, but not necessarily malicious.

explanation: It's like a guest who overstays their welcome - they're not necessarily bad, but they're not exactly helpful either.

real-world examples: Adware, spyware, and browser toolbars that collect user data or display unwanted advertisements.

related terms: Malware, Potentially Unwanted Program (PUP), Adware

Green Hat Hacker

definition: A novice hacker learning the ropes of hacking and cybersecurity.

explanation: It's like an apprentice learning a trade.

real-world examples: Beginners experimenting with hacking tools and techniques to understand cybersecurity.

related terms: Hacker, Cybersecurity, White Hat Hacker

Group Policy

definition: A feature of Microsoft Windows that allows administrators to manage and configure settings for users and computers on a network.

explanation: It's like a set of rules that applies to everyone in a group, ensuring consistency and control.

real-world examples: Used to enforce security policies, configure software settings, and manage user access to resources.

related terms: Active Directory, Windows Server, Configuration Management

Hacker

definition: An individual who uses computer, networking, or other skills to overcome a technical problem.

explanation: In the cybersecurity world, a hacker is someone who explores and exploits vulnerabilities in systems and networks. Hackers can be categorized as "black hat" (malicious), "white hat" (ethical), or "gray hat" (somewhere in between).

real-world examples: A "black hat" hacker might steal credit card data, while a "white hat" hacker might work for a security company, testing systems for weaknesses.

related terms: Cracker, Ethical Hacker, Security Researcher

Hacker Ethics

definition: A set of moral principles that guide the behavior of hackers.

explanation: It's like a code of conduct for hackers, emphasizing the responsible use of their skills and the importance of sharing knowledge.

real-world examples: Some key principles include information should be free, mistrust authority, and computers can be used for good.

related terms: Hacker Culture, Information Freedom, Open-Source Software

Hacker Group

definition: A collective of hackers who work together to achieve a common goal.

explanation: Imagine a group of friends who love playing pranks but instead of toilet-papering houses, they're breaking into computer systems.

real-world examples: Anonymous, LulzSec, and Lizard Squad are examples of infamous hacker groups.

related terms: Hacking, Cybercrime, Activism

Hacktivist

definition: An individual or group that uses hacking to promote political ends, often through defacement or DDoS attacks.

explanation: It's like a protester using digital means to draw attention to their cause.

real-world examples: Groups like Anonymous carrying out attacks to promote free speech or protest government actions.

related terms: Activism, Cyberattack, Defacement

Hardening

definition: The process of securing a system by reducing its surface of vulnerability.

explanation: Think of it as reinforcing a castle's walls to make it more difficult for attackers to breach.

real-world examples: Disabling unnecessary services, applying security patches, and configuring firewalls are all examples of hardening techniques.

related terms: Security Configuration, Vulnerability Management, System Hardening

Hardware Firewall

definition: A physical device that regulates incoming and outgoing network traffic based on security rules.

explanation: It's like a security guard for your network, checking the ID of every data packet that tries to enter or leave.

real-world examples: Cisco ASA, Palo Alto Networks, and Fortinet FortiGate are popular hardware firewalls.

related terms: Firewall, Network Security, Perimeter Security

Hardware Keylogger

definition: A physical device that records keystrokes on a computer keyboard.

explanation: It's like a tiny spy hidden inside your keyboard, secretly recording everything you type.

real-world examples: Often used for malicious purposes, such as stealing passwords or credit card information.

related terms: Keylogger, Spyware, Surveillance

Hardware Security Module (HSM)

definition: A physical device that safeguards and manages digital keys for strong authentication and provides crypto processing.

explanation: Imagine a high-security vault for your digital keys, protecting them from theft and unauthorized use.

real-world examples: Used to secure financial transactions, generate, and manage cryptographic keys, and protect sensitive data.

related terms: Cryptography, Encryption, Key Management

Hash

definition: A fixed-size string of characters generated from a variable-length input using a mathematical algorithm.

explanation: It's like a digital fingerprint for data - a unique identifier that can be used to verify its integrity.

real-world examples: Used to store passwords securely, verify file integrity, and ensure data hasn't been tampered with.

related terms: Hash Function, Checksum, Cryptography

Hash Collision

definition: A situation where two different inputs generate the same hash value.

explanation: It's like two different people having the same fingerprints - it's incredibly rare but not impossible.

real-world examples: Collision attacks can be used to break cryptographic hash functions and create forged digital signatures.

related terms: Hash Function, Collision Attack, Cryptography

Hash Function

definition: A mathematical algorithm that transforms an input (or 'message') into a fixed-size string of bytes.

explanation: It's like a meat grinder that turns various ingredients into a uniform sausage. The output (the hash) is always the same length, regardless of the size or complexity of the input.

real-world examples: Used to create digital signatures, verify file integrity, and store passwords securely.

related terms: Hash, Checksum, Cryptography

Hashcat

definition: A popular password recovery tool that cracks password hashes using brute-force and dictionary attacks.

explanation: It's like a locksmith's toolkit for cracking passwords.

real-world examples: Used by security professionals for penetration testing and malicious actors to gain unauthorized access to accounts.

related terms: Password Cracking, Brute Force Attack, Dictionary Attack

Hashing

definition: The process of converting data into a fixed-size string of characters, which is typically a hash code.

explanation: It's like creating a unique fingerprint for a piece of data.

real-world examples: Storing hashed passwords in a database to prevent easy access if the database is breached.

related terms: Encryption, Cryptography, Data Integrity

Header

definition: The portion of a data packet that contains metadata about the packet, such as its source and destination addresses, protocol type, and other control information.

explanation: It's like the address label on a package, telling the postal service where to deliver it.

real-world examples: Network headers contain information like the sender's IP address, the recipient's IP address, and the type of data being transmitted.

related terms: Data Packet, Network Protocol, IP Header

Header Manipulation

definition: Modifying the header information in a data packet to deceive or mislead network devices or security systems.

explanation: It's like changing the address on a letter to trick the postal service into delivering it to the wrong person.

real-world examples: Can bypass firewalls, impersonate other users, or redirect traffic to malicious websites.

related terms: Packet Forgery, Spoofing, Man-in-the-Middle Attack

Health Insurance Portability and Accountability Act (HIPAA)

definition: A U.S. federal law designed to protect patient health information from being disclosed without the patient's consent or knowledge.

explanation: It's like a security blanket ensuring medical information stays private and secure.

real-world examples: Healthcare providers implementing safeguards to protect patient data.

related terms: Data Privacy, Regulatory Compliance, Healthcare

Heartbleed

definition: A security vulnerability in the OpenSSL cryptographic software library.

explanation: It's like a leaky heart - the vulnerability exposed sensitive information from the memory of systems using OpenSSL.

real-world examples: The Heartbleed bug was discovered in 2014, allowing attackers to steal private keys, login credentials, and other confidential data.

related terms: Vulnerability, OpenSSL, Security Bug

Heartbleed Bug

definition: A synonym for Heartbleed.

explanation: This refers to the same vulnerability in the OpenSSL cryptographic software library.

real-world examples: See Heartbleed.

related terms: Vulnerability, OpenSSL, Security Bug

Heuristic Analysis

definition: A method of analyzing files or behavior to detect potential threats based on patterns or characteristics.

explanation: It's like a detective profiling a suspect based on their behavior and appearance.

real-world examples: Used by antivirus software to detect new or unknown malware that doesn't match known signatures.

related terms: Malware Detection, Antivirus, Signature-Based Detection

Heuristic Antivirus

definition: Antivirus software that uses heuristic analysis to detect malware.

explanation: It's like a detective with a keen eye for detail, spotting suspicious behavior even when it doesn't match known patterns.

real-world examples: More effective at detecting new or unknown malware than signature-based antivirus software but can produce more false positives.

related terms: Antivirus, Malware Detection, Heuristic Analysis

Hidden Field

definition: A form field in a web page that is not visible to the user but can still be accessed and manipulated by an attacker.

explanation: It's like a hidden compartment in a piece of furniture that the owner doesn't know about.

real-world examples: Attackers can use hidden fields to inject malicious data or bypass security controls.

related terms: Web Application Security, Cross-Site Request Forgery (CSRF), Input Validation

Hidden File

definition: A file not displayed by default in a file manager or directory listing.

explanation: It's like a secret room in a house hidden behind a bookshelf.

real-world examples: Operating systems and applications often use hidden files to store configuration settings or other data that the user doesn't need to see.

related terms: File System, Operating System, Privacy

High Assurance

definition: A level of confidence that a system meets its security requirements and is free from vulnerabilities.

explanation: Imagine a bank vault with multiple layers of security, including steel doors, biometric scanners, and armed guards. A high-assurance system is like that vault -designed to be extremely secure and resistant to attack.

real-world examples: Military-grade systems, critical infrastructure control systems, and high-security financial systems often require high assurance.

related terms: Security Engineering, Trusted Systems, Formal Verification

High Availability (HA)

definition: The ability of a system or component to operate continuously without interruption for a long time.

explanation: Imagine a hospital's power supply that never fails, even during a blackout. High availability ensures that critical systems are always up and running, even in the face of failures or attacks.

real-world examples: Redundant servers, failover mechanisms, and load balancing are all used to achieve high availability.

related terms: Disaster Recovery, Fault Tolerance, Business Continuity

Hijacking

definition: Taking control of a communication session, such as a TCP connection or a browser session.

explanation: It's like a car thief taking control of your vehicle while you're driving.

real-world examples: Session hijacking, browser hijacking, and DNS hijacking are all examples of hijacking attacks.

related terms: Session Hijacking, Man-in-the-Middle Attack, Cyber Attack

HIPAA

definition: A U.S. federal law that sets standards for protecting sensitive patient health information.

explanation: It's like a set of rules for doctors and hospitals on handling patient data responsibly.

real-world examples: HIPAA requires healthcare providers to implement safeguards to protect patient information from unauthorized access, disclosure, alteration, or destruction.

related terms: PHI (Protected Health Information), Healthcare IT, Compliance

Honeynet

definition: A network of honeypots designed to attract and analyze attacker activity.

explanation: It's like a trap set for a burglar – the honeynet appears to be a vulnerable system. Still, it's a decoy allowing security professionals to monitor and learn from attacker behavior.

real-world examples: Used to gather intelligence on attacker tactics, techniques, and procedures (TTPs) and to develop better defenses.

related terms: Honeypot, Deception Technology, Threat Intelligence

Honeypot

definition: A decoy system or resource designed to attract and deceive attackers.

explanation: It's like a fake security camera that's recording a burglar's every move.

real-world examples: Honeypots can distract attackers from real systems, gather information about their tactics, and even trap them in a controlled environment.

related terms: Deception Technology, Cyber Threat Hunting, Intrusion Detection System (IDS)

Host Intrusion Prevention System (HIPS)

definition: A security software that monitors and analyzes the behavior of processes and applications running on a host system to detect and prevent malicious activity.

explanation: It's like a security guard who patrols the inside of a building, looking for suspicious activity and stopping it before it can cause harm.

real-world examples: HIPS can detect and block malware, unauthorized software installations, and other types of attacks.

related terms: Intrusion Prevention System (IPS), Endpoint Security, Host-Based Security

Host-Based Firewall (HBF)

definition: A firewall installed and running on an individual host system (e.g., a laptop or server) to protect it from network-based attacks.

explanation: It's like a personal bodyguard for your computer, blocking unwanted traffic and protecting against intrusions.

real-world examples: Windows Firewall and macOS Firewall are examples of host-based firewalls.

related terms: Firewall, Network Security, Endpoint Security

Host-based Intrusion Detection System (HIDS)

definition: A security software monitoring the characteristics of a single host and the events occurring within that host for suspicious activity.

explanation: It's like a burglar alarm for your computer, alerting you if someone tries to break in or if something unusual is happening.

real-world examples: HIDS can detect changes to system files, unauthorized access attempts, and other signs of compromise.

related terms: Intrusion Detection System (IDS), Endpoint Security, Log Analysis

Hot Site

definition: A backup facility equipped with hardware, software, and data and can be quickly activated in the event of a disaster.

explanation: It's like a fully furnished spare house that you can move into immediately if your primary home is destroyed.

real-world examples: Used for disaster recovery when a primary site is unavailable, allowing for minimal downtime and data loss.

related terms: Disaster Recovery, Cold Site, Warm Site

Hotfix

definition: A software update that addresses a specific bug or vulnerability.

explanation: It's like a band-aid for a software problem, providing a quick fix until a more permanent solution can be implemented.

real-world examples: Microsoft releases security hotfixes on Patch Tuesday to address critical vulnerabilities in its software.

related terms: Patch, Security Update, Software Update

HTML Injection

definition: An attack that injects malicious HTML code into a vulnerable web page.

explanation: It's like a prankster adding graffiti to a website, altering its appearance or functionality.

real-world examples: Attackers can use HTML injection to deface websites, steal sensitive information, or redirect users to malicious websites.

related terms: Injection Attack, Cross-Site Scripting (XSS), Web Application Security

HTTP Response Splitting

definition: An attack that exploits vulnerabilities in HTTP headers to inject malicious content.

explanation: It's like splitting a message in two to manipulate its meaning.

real-world examples: Injecting malicious scripts through HTTP response headers.

related terms: Web Application Security, Injection Attacks, Cybersecurity

HTTP Strict Transport Security (HSTS)

definition: A security mechanism that tells browsers to only communicate with a website over HTTPS, even if the user tries to access it over HTTP.

explanation: It's like a sign on a door that says "No Trespassing" - it tells browsers that they are not allowed to access the website over an insecure connection.

real-world examples: HSTS helps to prevent man-in-the-middle (MitM) attacks and other forms of eavesdropping.

related terms: HTTPS, Web Security, Man-in-the-Middle Attack

HTTPS

definition: The secure version of HTTP, the protocol to transfer data over the web.

explanation: It's like sending a letter in a sealed envelope - the contents are encrypted and cannot be read by anyone who intercepts it.

real-world examples: Used to secure online transactions, protect user privacy, and prevent eavesdropping.

related terms: SSL/TLS, Encryption, Web Security

Human Firewall

definition: A person trained to recognize and respond to social engineering attacks and other security threats.

explanation: They are like the human equivalent of a firewall, acting as the last line of defense against social engineering attacks.

real-world examples: Educating employees about phishing scams, suspicious emails, and other social engineering tactics.

related terms: Security Awareness Training, Social Engineering, Phishing

Human-Computer Interaction (HCI)

definition: The study of how people interact with computers and how to design computer systems that are easy to use and understand.

explanation: It's like designing a user-friendly smartphone or video game interface.

real-world examples: HCI principles are applied to the security software design, making it easier for users to understand and follow security best practices.

related terms: User Interface (UI) Design, Usability, User Experience (UX)

Hybrid Analysis

definition: A method of malware analysis that combines static and dynamic analysis techniques.

explanation: It's like a doctor using both a physical exam and lab tests to diagnose a patient.

real-world examples: Analyzing the code of a malware sample (static analysis) and observing its behavior in a sandbox environment (dynamic analysis).

related terms: Malware Analysis, Static Analysis, Dynamic Analysis

Hybrid Attack

definition: An attack combining multiple techniques, such as phishing and malware, to increase effectiveness.

explanation: It's like a burglar using a crowbar and a lockpick to break into a house.

real-world examples: A phishing email containing a link to a malicious website exploits a vulnerability in the user's browser.

related terms: Phishing, Malware, Vulnerability

Hybrid Cloud

definition: A cloud computing environment that combines public and private clouds.

explanation: It's like having a house with some rooms that are open to the public (public cloud) and others that are private (private cloud).

real-world examples: A company might use a public cloud for non-sensitive data and a private cloud for sensitive data.

related terms: Cloud Computing, Public Cloud, Private Cloud

Hybrid Cloud Security

definition: The practice of securing hybrid cloud environments.

explanation: It's like having a security system that protects your public and private rooms.

real-world examples: Implementing security controls across both public and private cloud environments to ensure consistent protection.

related terms: Cloud Security, Public Cloud Security, Private Cloud Security

Hybrid Encryption

definition: A method that connects symmetric and asymmetric encryption to achieve speed and security.

explanation: Imagine sending a package with a combination lock (symmetric encryption for speed) but sending the combination separately in a sealed envelope (asymmetric encryption for secure key exchange).

real-world examples: Commonly used in SSL/TLS for secure web browsing.

related terms: Symmetric Encryption, Asymmetric Encryption, SSL/TLS

Hybrid Warfare

definition: A military strategy that blends conventional warfare with cyber warfare, disinformation, and other non-military tactics.

explanation: It's like a chess game where you use your pieces and psychological tricks to outmaneuver your opponent.

real-world examples: Russia's interference in the 2016 U.S. elections is an example of hybrid warfare.

related terms: Cyber Warfare, Information Warfare, Disinformation

Hyperjacking

definition: A type of attack that compromises the hypervisor, the software that manages virtual machines.

explanation: It's like taking over the control tower at an airport, allowing the attacker to manipulate or monitor all the virtual machines running on the host.

real-world examples: An attacker could access all virtual machines on a compromised server, potentially stealing data or causing disruption.

related terms: Virtual Machine (VM), Hypervisor, Virtualization Security

Hypertext Preprocessor (PHP)

definition: A popular server-side scripting language used for web development.

explanation: It's like the engine that powers many websites, allowing them to dynamically generate content and interact with databases.

real-world examples: PHP is used by popular content management systems like WordPress and Drupal.

related terms: Web Development, Server-Side Scripting, Web Application Security

Hypertext Transfer Protocol (HTTP)

definition: The protocol used to transfer data over the web.

explanation: It's like the language that web browsers and servers use to communicate with each other.

real-world examples: Every time you visit a website, your browser sends HTTP requests to the server, and the server responds with HTTP responses.

related terms: Web Browser, Web Server, HTTPS

Hypertext Transfer Protocol Secure (HTTPS)

definition: The secure version of HTTP, which uses encryption to protect data in transit.

explanation: It's like sending a letter in a sealed envelope – the contents are encrypted and cannot be read by anyone who intercepts it.

real-world examples: Used for secure online transactions, such as online banking and shopping.

related terms: HTTP, Encryption, SSL/TLS

Hypervisor

definition: Software that creates and runs virtual machines (VMs).

explanation: Think of it as a landlord who rents out virtual apartments (VMs) on a single physical computer.

real-world examples: Hypervisors like VMware ESXi, Microsoft Hyper-V, and KVM.

related terms: Virtual Machine (VM), Virtualization, Cloud Computing

I

ICMP Flood

definition: A type of denial-of-service (DoS) attack that floods a target with Internet Control Message Protocol (ICMP) packets.

explanation: It's like sending a barrage of text messages to someone's phone, overwhelming it and preventing legitimate messages from getting through.

real-world examples: Ping flood and Smurf attack.

related terms: DoS Attack, DDoS Attack, Botnet

Identity and Access Management (IAM)

definition: The framework of policies and technologies facilitating the management of electronic or digital identities.

explanation: It's like a system of locks, keys, and security guards that control who can access what resources in a building.

real-world examples: User provisioning, access control, authentication, and authorization.

related terms: Identity Management, Access Control, Authentication

Identity as a Service (IDaaS)

definition: A cloud-based service that provides identity and access management (IAM) abilities.

explanation: It's like a cloud-based security guard service that handles authentication and authorization for you.

real-world examples: Okta, OneLogin, and Azure Active Directory.

related terms: Identity and Access Management (IAM), Cloud Computing, Single Sign-On (SSO)

Identity Federation

definition: The practice of linking a user's identity across multiple identity management systems.

explanation: It's like having a universal ID card that works at multiple organizations.

real-world examples: Single sign-on (SSO) solutions, where you use one set of credentials to access multiple applications.

related terms: Identity Management, Single Sign-On (SSO), Authentication

Identity Governance and Administration (IGA)

definition: The policies, processes, and technologies used to manage user identities and access rights throughout their lifecycle.

explanation: It's like a comprehensive system for managing employee IDs, from the moment they're hired to the moment they leave the company.

real-world examples: User provisioning, deprovisioning, access reviews, and password management.

related terms: Identity Management, Access Management, Governance, Risk Management, and Compliance (GRC)

Identity Management

definition: The organizational processes and technologies utilized to identify, authenticate, and authorize individuals or groups of people to access applications, systems, or networks.

explanation: It's like a system that keeps track of who everyone is and what they're allowed to do in a digital environment.

real-world examples: Usernames, passwords, biometrics, and access control lists.

related terms: Authentication, Authorization, Access Management

Identity Proofing

definition: The process of verifying that a user is who they claim to be.

explanation: It's like showing your ID to a bouncer at a club.

real-world examples: Asking for a government-issued ID, verifying an email address, or using knowledge-based authentication (KBA) questions.

related terms: Authentication, Identity Verification, Know Your Customer (KYC)

Identity Provider (IdP)

definition: A system entity that can create, maintain, and manage entity information for principals and provide authentication services to relying party applications within a federation or distributed network.

explanation: It's the gatekeeper of your digital identity, verifying who you are and issuing credentials that can be used to access other systems.

real-world examples: Google, Facebook, and Twitter can act as identity providers when you use their accounts to log into other websites.

related terms: Identity Management, Single Sign-On (SSO), Federation

Identity Theft

definition: The fraudulent acquisition and use of a person's private identifying information, usually for financial gain.

explanation: It's like someone stealing your wallet and using your credit cards and personal information to make purchases or open new accounts in your name.

real-world examples: Stealing credit card numbers, social security numbers, or other personal information to open fraudulent accounts or make unauthorized purchases.

related terms: Fraud, Cybercrime, Social Engineering

Image Analysis

definition: The process of extracting information from images using computer vision techniques.

explanation: It's like teaching a computer to see and understand the content of an image.

real-world examples: Facial recognition, object detection, and image classification.

related terms: Computer Vision, Artificial Intelligence (AI), Machine Learning

Image Spam

definition: Spam that uses images to bypass traditional text-based spam filters.

explanation: It's like a spammer trying to disguise their message as a picture postcard.

real-world examples: Image spam often contains embedded text or links to malicious websites.

related terms: Spam, Phishing, Malware

Impersonation Attack

definition: A type of attack where an attacker impersonates a trusted entity to gain unauthorized access or information.

explanation: It's like a con artist who poses as a bank employee to trick you into giving them your account information.

real-world examples: Phishing emails, vishing (voice phishing) calls, and social engineering attacks often involve impersonation.

related terms: Social Engineering, Phishing, Identity Theft

Implicit Deny

definition: A security principle where anything that is not explicitly allowed is automatically denied.

explanation: Think of it like a picky eater who only eats foods on their approved list. Anything else is automatically rejected.

real-world examples: In firewalls, if a rule doesn't explicitly allow traffic, it's blocked.

related terms: Access Control, Firewall Rules, Security Policy

Incident

definition: An event that could negatively affect the confidentiality, integrity, or availability of an organization's assets.

explanation: It's like a fire alarm going off - it signals that something might be wrong and needs investigation.

real-world examples: A malware infection, a data breach, or a system outage are all incidents.

related terms: Incident Response, Security Event, Breach

Incident Management

definition: Managing security incidents from detection to resolution.

explanation: It's like a fire drill - you practice what to do in case of a fire, so you're prepared to act quickly and effectively if it happens.

real-world examples: Identifying, analyzing, containing, eradicating, and recovering from a security incident.

related terms: Incident Response, Incident Response Plan, Security Operations Center (SOC)

Incident Response

definition: The process of responding to and addressing the aftermath of a security breach or cyberattack (also known as an incident).

explanation: It's like calling 911 and having emergency responders show up to help you.

real-world examples: Identifying the cause of the incident, containing the damage, eradicating the threat, and recovering normal operations.

related terms: Incident Management, Incident Response Plan, Computer Security Incident Response Team (CSIRT)

Incident Response Plan

definition: A documented plan that outlines the procedures an organization will follow when responding to a security incident.

explanation: It's like a fire escape plan for a building - it tells you what to do in case of emergency so you can get to safety.

real-world examples: A step-by-step guide for identifying, containing, eradicating, and recovering from a security incident.

related terms: Incident Response, Incident Management, Disaster Recovery Plan (DRP)

Incident Response Team (IRT)

definition: A group of individuals responsible for handling and responding to security incidents.

explanation: They are the cybersecurity first responders, trained to deal with cyber attacks and minimize damage.

real-world examples: A team of security analysts, forensic investigators, and IT professionals who work together to respond to a security incident.

related terms: Incident Response, Computer Security Incident Response Team (CSIRT), Security Operations Center (SOC)

Indicator of Attack (IoA)

definition: Observable events or artifacts that signify an ongoing attack or breach.

explanation: It's like seeing signs that a burglar is trying to break in, such as a broken window or a tripped alarm.

real-world examples: Detecting unusual network traffic patterns or unauthorized access attempts that indicate a cyberattack.

related terms: Indicator of Compromise (IoC), Threat Detection, Incident Response, Security Monitoring

Indicators of Compromise (IoCs)

definition: Artifacts or evidence found on a network or in an operating system that, with high confidence, indicates a computer intrusion.

explanation: These are like clues left behind by a burglar – they can help you identify what happened and who was responsible.

real-world examples: Unusual network traffic patterns, unexpected file modifications, or the presence of malware.

related terms: Threat Intelligence, Incident Response, Security Monitoring

Industrial Control System (ICS)

definition: A system used to control industrial processes, such as those found in power plants, factories, and transportation systems.

explanation: Think of it like the brain of a factory, controlling all the machines and processes that keep things running.

real-world examples: Programmable logic controllers (PLCs), supervisory control and data acquisition (SCADA) systems, and distributed control systems (DCS).

related terms: Critical Infrastructure, Operational Technology (OT), Industrial Cybersecurity

Industrial Espionage

definition: The theft of trade secrets or confidential information from a company or organization.

explanation: It's like a rival company stealing your secret recipe or business plan.

real-world examples: Attackers might use cyber espionage to steal proprietary information, gain a competitive advantage, or disrupt operations.

related terms: Cyber Espionage, Intellectual Property Theft, Trade Secret Theft

Industrial IoT Attacks

definition: Cyberattacks targeting industrial internet of things (IoT) devices and networks.

explanation: It's like sabotaging factory equipment to disrupt production.

real-world examples: Attacks on SCADA systems controlling critical infrastructure.

related terms: IoT Security, Cybersecurity, Critical Infrastructure

Information Assurance (IA)

definition: The practice of assuring information and managing risks related to the use, processing, storage, and transmission of information.

explanation: It's like a bodyguard for your information, protecting it from unauthorized access, disclosure, alteration, or destruction.

real-world examples: Implementing security controls, monitoring systems for threats, and responding to incidents.

related terms: Cybersecurity, Information Security, Risk Management

Information Leakage

definition: The unauthorized disclosure of confidential information.

explanation: It's like a leak in a pipe – sensitive information is slowly dripping out of the organization without anyone noticing.

real-world examples: Employees accidentally emailing confidential documents to the wrong person, or a company misconfiguring a cloud server, making data accessible to the public.

related terms: Data Breach, Data Loss, Insider Threat

Information Rights Management (IRM)

definition: A technology that protects sensitive information by controlling how it can be accessed, used, and shared.

explanation: It's like a digital lock on a file that only authorized users can open.

real-world examples: Preventing employees from forwarding confidential emails or printing sensitive documents.

related terms: Data Loss Prevention (DLP), Digital Rights Management (DRM), Access Control

Information Security

definition: The practice of safeguarding information from unauthorized access, use, disclosure, disruption, modification, or destruction.

explanation: It's like a security system for your data, safeguarding it from theft, damage, or loss.

real-world examples: Implementing firewalls, antivirus software, encryption, and strong passwords.

related terms: Cybersecurity, Information Assurance (IA), Data Security

Information Security Management

definition: The process of protecting an organization's information assets from threats.

explanation: It's like having a robust security system to protect valuable data.

real-world examples: Implementing ISO/IEC 27001 for information security management.

related terms: Cybersecurity, Risk Management, Compliance

Information Security Management System (ISMS)

definition: A systematic approach to managing sensitive company information so that it remains secure.

explanation: It's like a comprehensive security plan for an organization's information, covering everything from risk assessment to incident response.

real-world examples: ISO 27001 is a widely used standard for implementing an ISMS.

related terms: Information Security, Cybersecurity, Risk Management

Information Security Policy

definition: A set of rules and procedures that define how an organization protects its information assets.

explanation: It's like a rulebook for employees on handling sensitive data, using company computers, and accessing the network.

real-world examples: Defining acceptable use policies, password policies, and incident response procedures.

related terms: Information Security, Cybersecurity, Compliance

Information Sharing and Analysis Center (ISAC)

definition: A collaborative organization that provides a central resource for gathering and sharing information on cyber threats.

explanation: It's like a neighborhood watch group for cybersecurity.

real-world examples: Financial services ISAC sharing threat intelligence among banks.

related terms: Threat Intelligence, Cybersecurity, Collaboration

Information Warfare

definition: The use of information and information technology to gain an advantage over adversary.

explanation: It's like a battle fought with information instead of weapons.

real-world examples: Spreading propaganda, manipulating public opinion, and disrupting communication networks.

related terms: Cyber Warfare, Psychological Warfare, Disinformation

InfoSec

definition: Short for Information Security, referring to the practice of protecting information by mitigating information risks.

explanation: It's like putting locks and alarms on valuable information to keep it safe.

real-world examples: Implementing firewalls, encryption, and access controls to protect data.

related terms: Cybersecurity, Data Protection, IT Security

Infrastructure as a Service (IaaS)

definition: A cloud computing model that provides virtualized computing resources over the internet.

explanation: It's like renting a computer in the cloud, instead of buying and maintaining your hardware.

real-world examples: Amazon Web Services (AWS), Microsoft Azure, and Google Cloud Platform (GCP) are popular IaaS providers.

related terms: Cloud Computing, Cloud Service Provider (CSP), Platform as a Service (PaaS)

Infrastructure as Code (IaC)

definition: The practice of managing and provisioning infrastructure through machine-readable definition files, rather than physical hardware configuration or interactive configuration tools.

explanation: It's like using a blueprint to build a house instead of manually laying each brick. IaC allows you to automate the provisioning and management of infrastructure resources.

real-world examples: Tools like Terraform, Ansible, and AWS CloudFormation are used to implement IaC.

related terms: Cloud Computing, DevOps, Automation

Injection Attack

definition: An attack that exploits a vulnerability in a web application by injecting malicious code or data into a query or command.

explanation: It's like sneaking extra ingredients into a recipe, changing the intended outcome.

real-world examples: SQL injection, command injection, and cross-site scripting (XSS).

related terms: Web Application Security, Vulnerability, Exploit

Injection Vulnerability

definition: A weakness in a web application that lets an attacker inject malicious code or data into a query or command.

explanation: It's like a hole in a fence that allows an intruder to enter a property.

real-world examples: Unvalidated input fields, improper handling of user-supplied data, and lack of output encoding.

related terms: Injection Attack, Web Application Security, Vulnerability

Input Sanitization

definition: The process of cleaning or filtering user input to prevent it from being used maliciously.

explanation: It's like washing your hands before dinner to remove germs that could make you sick.

real-world examples: Removing special characters from user input to prevent SQL injection attacks or encoding HTML tags to prevent cross-site scripting (XSS) attacks.

related terms: Input Validation, Injection Attacks, Web Application Security

Input Validation

definition: The process of checking user input to ensure it conforms to expected values and formats.

explanation: It's like a club bouncer checking IDs to ensure everyone is of legal age.

real-world examples: Checking that a user-entered phone number contains only digits and is the correct length.

related terms: Input Sanitization, Injection Attacks, Web Application Security

Insecure Deserialization

definition: A vulnerability that arises when an application deserializes (reconstructs) untrusted data without proper validation.

explanation: It's like opening a mystery package from a stranger without checking what's inside.

real-world examples: Attackers can exploit insecure deserialization to inject malicious code into a system, leading to remote code execution and data breaches.

related terms: Serialization, Deserialization, Injection Attacks

Insecure Direct Object Reference (IDOR)

definition: A vulnerability that occurs when an application exposes a direct reference to an internal object, such as a file or database record, that allows an attacker to manipulate it.

explanation: It's like a bank teller giving you access to someone else's account just because you asked for it.

real-world examples: An attacker could change their user ID in a URL to view another user's order history or access their personal information.

related terms: Access Control, Authorization, Vulnerability

Insider

definition: A person with authorized access to an organization's systems and data, such as an employee, contractor, or partner.

explanation: Insiders can be a significant source of security risks, either intentionally or unintentionally.

real-world examples: An employee who steals company secrets to sell to a competitor or a contractor who accidentally deletes critical files.

related terms: Insider Threat, Insider Attack, Data Loss Prevention (DLP)

Insider Attack

definition: A malicious act carried out by an insider against their organization.

explanation: It's like a wolf in sheep's clothing – an insider who abuses their trusted access to steal data, sabotage systems, or cause other harm.

real-world examples: A disgruntled employee leaking confidential information or a contractor planting malware on the company's network.

related terms: Insider Threat, Sabotage, Data Breach

Insider Threat

definition: The potential for an insider to harm an organization through sabotage, theft, fraud, or the disclosure of confidential information.

explanation: It's the risk that someone you trust could betray that trust and use their access to cause harm.

real-world examples: A disgruntled employee, a contractor seeking financial gain, or an individual an external attacker has compromised.

related terms: Insider Attack, Insider Threat Program, Security Awareness Training

Insider Threat Program

definition: A comprehensive program designed to detect, deter, and mitigate insider threats.

explanation: It's like a neighborhood watch program for the workplace, where everyone is vigilant about suspicious activity and reports it to the authorities.

real-world examples: Background checks, security awareness training, access controls, and monitoring tools.

related terms: Insider Threat, Security Awareness Training, User Behavior Analytics (UBA)

Integrated Development Environment (IDE)

definition: A software application that gives comprehensive facilities to computer programmers for software development.

explanation: It's like a toolbox for software developers, providing everything they need to write, test, and debug code in one place.

real-world examples: Microsoft Visual Studio, Eclipse, IntelliJ IDEA.

related terms: Software Development, Programming, Code Editor

Integrity

definition: The assurance that data has not been modified, altered, or corrupted unauthorizedly.

explanation: It's like ensuring that the original recipe for a dish hasn't been tampered with, so the final product tastes as intended.

real-world examples: Using checksums or hash functions to verify file integrity, implementing access controls to prevent unauthorized modifications, and using version control systems to track changes to code.

related terms: Data Integrity, Confidentiality, Availability

Integrity Measurement

definition: The process of verifying the integrity of a system or data by comparing its current state to a known good baseline.

explanation: It's like checking the fingerprints of a document to ensure it hasn't been altered.

real-world examples: Using hash functions to compare the current state of a file to a previously recorded hash value.

related terms: Data Integrity, Checksum, Hash Function

Intellectual Property (IP) Theft

definition: The unauthorized use or reproduction of intellectual property, like patents, trademarks, copyrights, or trade secrets.

explanation: It's like stealing someone else's ideas, inventions, or creative works.

real-world examples: Software piracy, counterfeit goods, and industrial espionage.

related terms: Intellectual Property Rights (IPR), Copyright Infringement, Patent Infringement

Intellectual Property Rights (IPR)

definition: Legal rights that protect creations of the mind, such as inventions, literary and artistic works, designs, symbols, names, and images used in commerce.

explanation: These are the legal protections given to creators for their original works, like books, music, and inventions.

real-world examples: Patents, trademarks, copyrights, and trade secrets.

related terms: Intellectual Property, Copyright, Patent

Interception Proxy

definition: A proxy server that sits between a client and server, allowing it to intercept and modify traffic.

explanation: It's like a middleman who intercepts your mail, reads it, and then delivers it to the intended recipient.

real-world examples: Used for web filtering, content caching, and security monitoring.

related terms: Proxy Server, Web Proxy, Man-in-the-Middle Attack (MITM)

Internal Audit

definition: An independent, objective assurance and consulting activity intended to add value and improve an organization's operations.

explanation: It's like an internal check-up to ensure everything is running smoothly and efficiently.

real-world examples: Auditing internal financial controls and processes.

related terms: Compliance, Risk Management, External Audit

Internal Controls

definition: Processes and procedures enforced to ensure the integrity of financial and accounting information, promote accountability, and prevent fraud.

explanation: It's like the safety measures put in place to ensure a machine runs safely and efficiently.

real-world examples: Segregation of duties and regular reconciliations.

related terms: Compliance, Risk Management, Fraud Detection

Internal Network Penetration Test

definition: A security assessment of an organization's internal network from an attacker's perspective.

explanation: It's like a security guard trying to break into a building to test its security.

real-world examples: Ethical hackers simulate real-world attacks to identify vulnerabilities in the internal network.

related terms: Penetration Testing, Vulnerability Assessment, Red Teaming

Internal Vulnerability Scan

definition: A security assessment that scans an organization's internal systems and networks for vulnerabilities.

explanation: It's like a health checkup for your internal network, identifying any weaknesses that attackers could exploit.

real-world examples: Using automated tools to scan for open ports, outdated software, and misconfigurations on internal systems.

related terms: Vulnerability Assessment, Penetration Testing, Security Scanning

Internet Control Message Protocol (ICMP)

definition: A network protocol utilized by network devices in order to send error messages and operational information.

explanation: It's like the postal service sending you a notification that your package couldn't be delivered.

real-world examples: Used for troubleshooting network problems but can also be exploited in attacks like ping floods.

related terms: Network Protocol, Ping, Traceroute

Internet Engineering Task Force (IETF)

definition: An open standards organization developing and promoting internet standards.

explanation: It's like a group of experts who come together to agree on how the internet should work.

real-world examples: The IETF is responsible for developing standards like TCP/IP, HTTP, and DNS.

related terms: Internet Standards, Network Protocols, Request for Comments (RFC)

Internet Message Access Protocol (IMAP)

definition: A protocol for retrieving email messages from a mail server.

explanation: It's like a postal worker who delivers your mail to your home, allowing you to read and manage it on your own computer.

real-world examples: IMAP allows you to access your email from multiple devices and keeps your messages synchronized across them.

related terms: Email Protocol, POP3, SMTP

Internet of Things (IoT)

definition: A network of physical devices embedded with electronics, software, sensors, actuators, and network connectivity enabling them to connect and exchange data.

explanation: It's like a network of smart devices that can communicate with each other and with the internet.

real-world examples: Smart thermostats, wearable fitness trackers, and connected appliances.

related terms: Smart Devices, Connected Devices, IoT Security

Internet Protocol (IP)

definition: A set of rules governing the format of data sent via the internet or local network.

explanation: Like a postal system's addressing format, IP ensures data packets reach their intended destinations.

real-world examples: The backbone of internet communication, assigning unique addresses (like 192.168.1.1) to devices.

related terms: TCP/IP, Ipv4, Ipv6

Internet Protocol (IP) Camera Hacking

definition: The unauthorized access and control of internet-connected cameras.

explanation: Think of it as a peeping Tom gaining access to your security cameras. Hackers can view live feeds, steal footage, or even use the camera to spy on you.

real-world examples: Exploiting vulnerabilities in camera firmware to gain access, using default passwords, or launching brute-force attacks.

related terms: IoT Security, Surveillance, Hacking

Internet Protocol Security (IPSec)

definition: A suite of protocols that provide authentication, integrity, and confidentiality for data transmitted over IP networks.

explanation: Like a secure courier service for your data packets, ensuring they arrive safely and haven't been tampered with.

real-world examples: Used in VPNs to encrypt traffic and ensure secure communication over the internet.

related terms: VPN, Encryption, Network Security

Internet Protocol Version 4 (Ipv4)

definition: The fourth version of the Internet Protocol (IP) that is used to identify devices on a network.

explanation: The older, more common IP addressing system, using four sets of numbers (e.g., 192.168.1.1).

real-world examples: Most devices currently use Ipv4, but it has a limited number of addresses, leading to the adoption of Ipv6.

related terms: IP Address, IP, Ipv6

Internet Protocol Version 6 (Ipv6)

definition: The most recent version of the Internet Protocol (IP) that is designed to address the limitations of Ipv4.

explanation: A newer, more expansive IP addressing system with a much larger address space than Ipv4.

real-world examples: Uses eight groups of hexadecimal digits (e.g. , 2001:db8:0:1234:0:567:8:1).

related terms: IP Address, IP, Ipv4

Internet Relay Chat (IRC)

definition: A protocol for real-time text-based communication.

explanation: Think of it like a giant chatroom where people from all over the world can communicate in real time.

real-world examples: Used for online communities, gaming, and even by some botnets for command and control.

related terms: Chat Protocol, Instant Messaging, Botnet

Internet Storm Center (ISC)

definition: A website that provides information about current internet security threats and attacks.

explanation: It's like a news feed for cyberattacks, keeping you informed about the latest threats.

real-world examples: Publishes daily security alerts, threat analyses, and other security-related information.

related terms: Threat Intelligence, Security News, Cyber Threat

Intrusion

definition: An unauthorized access to or use of a computer system or network.

explanation: Imagine someone breaking into your house – that's an intrusion.

real-world examples: A hacker gaining access to a company's servers, or a malware infection on a personal computer.

related terms: Cyber Attack, Vulnerability, Exploit

Intrusion Detection

definition: The process of monitoring systems and networks for signs of intrusion or unauthorized access.

explanation: It's like having a security camera that watches for intruders and alerts you if it sees anything suspicious.

real-world examples: Using intrusion detection systems (IDS) to monitor network traffic and system logs for signs of unauthorized activity.

related terms: Intrusion Detection System (IDS), Security Monitoring, Threat Detection

Intrusion Detection System (IDS)

definition: A software or hardware system that monitors network or system activities for malicious activity or policy violations and provides reports to a management station.

explanation: It's like a burglar alarm that goes off when it detects an intruder.

real-world examples: IDS can alert security personnel to suspicious activity, such as unauthorized access attempts or malware infections.

related terms: Intrusion Detection, Security Monitoring, Threat Detection

Intrusion Detection System Evasion

definition: Techniques used by attackers to avoid detection by intrusion detection systems (IDS).

explanation: It's like a burglar wearing a disguise to avoid being caught on camera.

real-world examples: Encrypting or obfuscating traffic, using fragmentation, or timing attacks.

related terms: IDS, Evasion, Cyber Attack

Intrusion Kill Chain

definition: A model that describes the phases of a cyber-attack, from reconnaissance to exfiltration.

explanation: It's like a playbook for hackers, outlining the steps they take to carry out an attack.

real-world examples: Understanding the kill chain can help defenders identify and stop attacks at earlier stages.

related terms: Threat Intelligence, Incident Response, Cyber Attack

Intrusion Prevention System (IPS)

definition: A network security appliance that monitors network or system activities for malicious activity or policy violations and produces reports to a management station.

explanation: It's like a security guard who not only monitors intruders but also tries to stop them from entering the building.

real-world examples: IPS can block malicious traffic, quarantine infected files, and take other actions to prevent attacks from succeeding.

related terms: Intrusion Detection System (IDS), Firewall, Network Security

Intrusion Signature

definition: A pattern of data that indicates a potential intrusion or attack.

explanation: It's like a fingerprint left at a crime scene that can be used to identify the attacker.

real-world examples: IDS and IPS use intrusion signatures to detect and block malicious activity.

related terms: Intrusion Detection System (IDS), Intrusion Prevention System (IPS), Signature-Based Detection

Intrusion Tolerance

definition: The ability of a system or network to continue operating even when it has been compromised.

explanation: It's like a ship that can stay afloat even with a hole in its hull.

real-world examples: Designing systems with redundancy and failover mechanisms so that they can continue operating even if some components are compromised.

related terms: Fault Tolerance, High Availability, Resilience

Intrusive Vulnerability Scan

definition: A vulnerability scan that actively tests for vulnerabilities by sending probes or exploits to a target system.

explanation: It's like a doctor poking and prodding a patient to diagnose an illness.

real-world examples: Can be more accurate than non-intrusive scans but also carries a higher risk of disrupting the target system.

related terms: Vulnerability Scan, Vulnerability Assessment, Penetration Testing

IoT Attacks

definition: Cyberattacks targeting internet of things (IoT) devices and networks.

explanation: It's like breaking into a smart home to control its devices.

real-world examples: Hacking smart thermostats to cause disruptions.

related terms: IoT Security, Cybersecurity, Network Security

IoT Security

definition: The practice of securing and protecting Internet of Things (IoT) devices and networks from cyber threats.

explanation: It's like installing a security system in your smart home to protect your connected devices from hackers.

real-world examples: Implementing strong passwords, keeping firmware up to date, and using network segmentation to isolate IoT devices from critical systems.

related terms: Internet of Things (IoT), Cybersecurity, Smart Device Security

IP Address

definition: A unique numerical label assigned to each device connected to a computer network that uses the Internet Protocol for communication.

explanation: It's like a street address for your computer or device on the internet.

real-world examples: Used to route traffic between devices on a network and identify them uniquely.

related terms: Ipv4, Ipv6, Network Address Translation (NAT)

IP Address Spoofing

definition: The generation of Internet Protocol (IP) packets with a false source IP address to hide the identity of the sender or impersonating another computing system.

explanation: It's like sending a letter with a fake return address to conceal your identity.

real-world examples: Used in denial-of-service (DoS) attacks, spam campaigns, and other malicious activities.

related terms: Spoofing, IP Spoofing, IP Address

IP Blacklisting

definition: The practice of blocking network traffic from specific IP addresses or ranges that are known to be malicious or associated with spam or other unwanted activity.

explanation: It's like a bouncer at a club who has a list of people who are not allowed inside.

real-world examples: Blocking IP addresses that are known to send spam emails or that have been used in cyberattacks.

related terms: Blacklist, IP Reputation, Spam Filtering

IP Fragmentation

definition: The process of dividing a large IP packet into smaller fragments to fit the maximum transmission unit (MTU) of a network.

explanation: Imagine sending a large package that needs to be split into smaller boxes to fit through a narrow doorway.

real-world examples: Used to transmit large packets over networks with smaller MTUs but can also be exploited in attacks.

related terms: IP Packet, MTU, IP Fragmentation Attack

IP Fragmentation Attack

definition: An attack that exploits the IP fragmentation process to evade security controls or overload a target system.

explanation: It's like sending a disassembled bomb through the mail, with the pieces arriving separately to avoid detection.

real-world examples: Teardrop attack, overlapping fragment attack.

related terms: DoS Attack, Packet Fragmentation, Security Evasion

IP Reputation

definition: A score assigned to an IP address based on its history of malicious activity.

explanation: It's like a credit score for an IP address – the higher the score, the more trustworthy it is.

real-world examples: Email filters and web application firewalls use IP reputation to block traffic from known malicious sources.

related terms: Blacklist, IP Blacklisting, Threat Intelligence

IP Spoofing

definition: The act of forging the source IP address in a network packet.

explanation: It is like sending a letter with a fake return address to make it appear to be from someone else.

real-world examples: Used in denial-of-service (DoS) attacks, spam campaigns, and other malicious activities.

related terms: Spoofing, IP Address Spoofing, Packet Forgery

IP Whitelisting

definition: A security measure that only allows traffic from specific IP addresses or ranges.

explanation: It is like having a VIP guest list for your network - only those on the list are allowed in.

real-world examples: Used to restrict access to sensitive systems or to limit the impact of attacks.

related terms: Blacklist, Firewall, Access Control

ISO/IEC 27001

definition: An international standard for managing information security.

explanation: It's like a gold standard for ensuring information security best practices are followed.

real-world examples: Organizations achieving ISO/IEC 27001 certification to demonstrate their commitment to information security.

related terms: Information Security, Cybersecurity, Compliance

IT Governance

definition: The framework for leadership, organizational structures, and processes to ensure that the organization's IT supports and enhances the achievement of its strategies and objectives.

explanation: It's like having a roadmap and rules for how technology is used within the organization.

real-world examples: Implementing COBIT for IT governance.

related terms: Corporate Governance, Risk Management, Compliance

J

Jailbreaking

definition: The process of removing software restrictions imposed by the manufacturer on a device, such as an iPhone or iPad.

explanation: It is like breaking out of jail – you remove the restrictions imposed by the device manufacturer to gain full control over the device.

real-world examples: Allows users to install unauthorized apps, customize the operating system, and access hidden features.

related terms: Rooting, iOS, Android

JavaScript Hijacking

definition: An attack that exploits vulnerabilities in JavaScript code to steal data or perform other malicious actions.

explanation: It is like a thief stealing your car keys while you're distracted.

real-world examples: Attackers can inject malicious JavaScript code into a website to steal cookies, redirect users, or launch other attacks.

related terms: Cross-Site Scripting (XSS), Web Application Security, JavaScript

JavaScript Object Notation (JSON)

definition: A lightweight data interchange format that is easy for humans to read and write and easy for machines to parse and create.

explanation: It is like a universal language for computers, allowing them to exchange data in a standardized format.

real-world examples: Used to transmit data between web applications and servers, and to store configuration settings.

related terms: Data Interchange Format, XML, REST API

K

Kerberos

definition: A network authentication protocol that uses secret-key cryptography to authenticate service requests between two hosts across an untrusted network.

explanation: It is like a complex system of tickets and passwords that allows users to access multiple services on a network without having to enter their credentials each time.

real-world examples: Commonly used in Windows networks to authenticate users and grant them access to resources.

related terms: Authentication, Single Sign-On (SSO), Active Directory

Key Derivation Function (KDF)

definition: A cryptographic hash function that derives one or more secret keys from a secret value, for example, a main key, a password, or a passphrase.

explanation: It is like a recipe for turning a simple password into a complex, difficult-to-crack key.

real-world examples: Used to strengthen weak passwords and to generate encryption keys from passwords.

related terms: Password Hashing, Encryption, Cryptography

Key Escrow

definition: A system in which a third party holds a copy of an encryption key.

explanation: It is like giving a spare key to a friend in case you lose yours.

real-world examples: Key escrow can be used to recover data if the original key is lost, but it also raises privacy concerns.

related terms: Encryption, Key Management, Cryptography

Key Exchange

definition: The process of sharing cryptographic keys between two or more parties in a secure manner.

explanation: It is like two spies exchanging secret codes in a crowded room without anyone else knowing.

real-world examples: Diffie-Hellman key exchange, RSA key exchange, and elliptic curve Diffie-Hellman (ECDH) key exchange.

related terms: Cryptography, Encryption, Public Key Infrastructure (PKI)

Key Management System (KMS)

definition: A system that automates the tasks of generating, storing, distributing, and revoking cryptographic keys.

explanation: It is like a high-tech key ring that keeps track of all your keys and makes it easy to manage them.

real-world examples: Used to manage encryption keys for data at rest, data in transit, and digital signatures.

related terms: Key Management, Encryption, Cryptography

Key Performance Indicators (KPIs)

definition: Quantifiable measures utilized in order to evaluate the success of an organization or of a particular activity in which it engages.

explanation: It's like the scorecard that shows how well the organization is performing.

real-world examples: Metrics like customer satisfaction scores and revenue growth rates.

related terms: Performance Measurement, Risk Management, Compliance

Key Risk Indicators (KRIs)

definition: Metrics utilized to provide an early signal of increasing risk exposures in various areas of an organization.

explanation: It's like a smoke detector that warns of potential fires.

real-world examples: Monitoring the number of cybersecurity incidents as an indicator of IT risk.

related terms: Risk Assessment, Risk Management, Compliance

Key Stretching

definition: A technique that enhances the security of weak passwords by repeatedly hashing them with salt.

explanation: It is like putting a weak password through a workout routine to make it stronger.

real-world examples: PBKDF2, bcrypt, and scrypt are common key stretching algorithms.

related terms: Password Hashing, Salt, Cryptography

Keylogger

definition: A kind of surveillance software that records every keystroke typed on a computer.

explanation: It is like having someone secretly looking over your shoulder as you type.

real-world examples: Not only used for malicious purposes, such as stealing passwords or credit card information, but can also be used for legitimate purposes like monitoring employee activity or parental control.

related terms: Spyware, Surveillance, Cybercrime

Keystroke Dynamics

definition: The analysis of a person's typing rhythm and patterns to authenticate their identity.

explanation: It is like recognizing someone by the way they walk – everyone has a unique typing style.

real-world examples: Used in biometric authentication systems to verify user identity.

related terms: Biometrics, Authentication, Behavioral Biometrics

Kill Chain

definition: A model that describes the phases of a cyber attack, from reconnaissance to exfiltration.

explanation: It is like a playbook for hackers, outlining the steps they take to carry out an attack.

real-world examples: Understanding the kill chain can help defenders identify and stop attacks at earlier stages.

related terms: Cyber Attack, Intrusion Kill Chain, Threat Intelligence

Know Your Customer (KYC)

definition: The process of verifying the identity of a client and assessing potential risks of illegal intentions for the business relationship.

explanation: It's like a bouncer checking IDs at a club to ensure only legitimate customers enter.

real-world examples: Banks requiring identification documents before opening accounts.

related terms: Anti-money laundering (AML), Compliance, Customer Due Diligence

Known Bad

definition: A file, IP address, domain, or another indicator that has been identified as malicious.

explanation: Think of it like a "Wanted" poster for cyber threats - if you see it, you know it is trouble.

real-world examples: Malware signatures, phishing URLs, and known bad IP addresses are all examples of known bads.

related terms: Blacklist, Threat Intelligence, Indicators of Compromise (IoCs)

Known Good

definition: A file, IP address, domain, or another indicator that has been verified as safe and trustworthy.

explanation: It is like a trusted friend you know you can rely on.

real-world examples: Whitelisting known good IP addresses or email addresses to allow them through a firewall.

related terms: Whitelist, Reputation, Trustworthy

Known Plaintext Attack

definition: A type of cryptanalysis attack where the attacker has access to both the plaintext and the corresponding ciphertext.

explanation: It is like trying to crack a code when you already know part of the message.

real-world examples: Attackers might use known plaintext attacks to try and deduce the encryption key or algorithm.

related terms: Cryptanalysis, Encryption, Ciphertext

L

Lateral Movement

definition: The techniques used by attackers to move through a network after gaining initial access.

explanation: It is like a burglar moving from room to room in a house, looking for valuables.

real-world examples: Attackers might use lateral movement to escalate privileges, access sensitive data, or install additional malware.

related terms: Privilege Escalation, Persistence, Network Security

LDAP (Lightweight Directory Access Protocol)

definition: A protocol for accessing and maintaining distributed directory information services over an Internet Protocol (IP) network.

explanation: Think of it as a phonebook for a computer network, storing information about users, groups, and other resources.

real-world examples: Active Directory, OpenLDAP, and other directory services use LDAP.

related terms: Active Directory, Authentication, Authorization

LDAP Injection

definition: An attack that injects malicious LDAP statements into a vulnerable application.

explanation: It is like a hacker sneaking their own commands into a database query, potentially allowing them to view or modify sensitive data.

real-world examples: Attackers can use LDAP injection to bypass authentication, gain unauthorized access, or modify data in a directory service.

related terms: Injection Attack, Web Application Security, LDAP

Least Privilege

definition: A security principle that states that a user, process, or program should only be granted the minimum privileges necessary to perform its function.

explanation: It is like giving a child a small allowance instead of a blank check - it limits the amount of damage they can do if they make a mistake.

real-world examples: Limiting user accounts to only the permissions they need to do their jobs or running applications with the lowest possible privileges.

related terms: Access Control, Privilege Management, Security Principle

Least Significant Bit (LSB) Steganography

definition: A steganography technique that hides data in the least significant bits of an image or audio file.

explanation: It is like hiding a message in the fine print of a document - It is there, but it is not immediately obvious.

real-world examples: Hiding text messages or images within seemingly innocent images or audio files.

related terms: Steganography, Data Hiding, Covert Channel

Legacy System

definition: An old technology, computer system, or application program, "of, relating to, or being a previous or outdated computer system."

explanation: It is like an old car that's still running, but it is not as efficient or secure as newer models.

real-world examples: Mainframe computers, outdated operating systems, and software applications that haven't been updated in years.

related terms: Legacy Software, Outdated Technology, Technical Debt

Lightweight Directory Access Protocol Secure (LDAPS)

definition: A secure version of the LDAP protocol that uses SSL/TLS to encrypt traffic.

explanation: It is like sending a letter in a sealed envelope – the contents are protected from prying eyes.

real-world examples: Used to securely access and manage directory information services over a network.

related terms: LDAP, Encryption, SSL/TLS

Lightweight Extensible Authentication Protocol (LEAP)

definition: A proprietary wireless LAN authentication method developed by Cisco Systems.

explanation: It is a now outdated and insecure authentication protocol for Wi-Fi networks, known for its vulnerabilities.

real-world examples: LEAP has been deprecated due to security weaknesses and should no longer be used.

related terms: Wireless Security, Authentication, Deprecated

Live CD

definition: A bootable optical disc or USB drive containing a complete operating system.

explanation: It is like a portable operating system that you can boot from without installing it on your hard drive.

real-world examples: Used for troubleshooting, data recovery, and forensic investigations.

related terms: Bootable Media, Linux Live CD, Forensic Toolkit

Load Balancer

definition: A device that distributes network or application traffic across multiple servers.

explanation: It is like a traffic cop directing cars to different lanes to avoid congestion.

real-world examples: Used to improve performance, scalability, and reliability of web applications and other services.

related terms: High Availability, Scalability, Network Traffic Management

Local Area Network (LAN)

definition: A computer network that connects computers within a limited area, in a residence, school, laboratory, university campus, or office building.

explanation: It is like a small neighborhood where computers can communicate with each other directly.

real-world examples: Home Wi-Fi networks, office networks, and school computer labs are all examples of LANs.

related terms: Network, Wide Area Network (WAN), Ethernet

Local File Inclusion (LFI)

definition: A type of web vulnerability that allows an attacker to include local files on a server.

explanation: It is like a burglar finding a key that unlocks a hidden room in a house.

real-world examples: Attackers can exploit LFI vulnerabilities to access sensitive files, configuration files, or even execute code on the server.

related terms: File Inclusion Vulnerability, Remote File Inclusion (RFI), Web Application Security

Lock Picking

definition: The practice of opening a lock without the original key.

explanation: It is like a skill that burglars use to bypass physical security measures.

real-world examples: Lock picking tools are used by locksmiths, security professionals, and, unfortunately, criminals.

related terms: Physical Security, Lock, Security Bypass

Lockout

definition: A security mechanism that temporarily or permanently disables a user account after a pre-defined number of failed login attempts.

explanation: It is like getting locked out of your house after entering the wrong password too many times.

real-world examples: Used to prevent brute-force attacks, where an attacker repeatedly tries to guess a password.

related terms: Account Lockout, Brute Force Attack, Authentication

Log Aggregation

definition: The process of gathering logs from multiple sources and centralizing them in a single location.

explanation: It is like gathering all the security camera footage from different parts of a building into a central monitoring room.

real-world examples: Log aggregation tools are used to collect and store logs from servers, applications, and network devices.

related terms: Log Management, Security Information and Event Management (SIEM), Log Analysis

Log Analysis

definition: The process of reviewing logs to identify security events, trends, or anomalies.

explanation: It is like a detective analyzing security camera footage to find clues about a crime.

real-world examples: Used to detect and investigate security incidents, monitor system performance, and troubleshoot problems.

related terms: Log Management, Security Information and Event Management (SIEM), Threat Detection

Log Management

definition: The process of collecting, storing, analyzing, and archiving logs.

explanation: It is like managing a library of security camera footage – you need to organize it, index it, and make it searchable.

real-world examples: Log management software is used to centralize logs from multiple sources, making it easier to analyze and investigate security incidents.

related terms: Log Analysis, Security Information and Event Management (SIEM), Event Log

Log Rotation

definition: The practice of archiving or deleting older log files to free up storage space and improve performance.

explanation: It is like cleaning out your closet – you get rid of old clothes to make room for new ones.

real-world examples: Configuring a server to automatically archive log files older than 30 days.

related terms: Log Management, System Logging, Data Retention

Log4j

definition: A popular Java-based logging library used by many applications and services.

explanation: It is like a notebook that records what happens in a software program – useful for debugging and troubleshooting.

real-world examples: A critical vulnerability in Log4j (Log4Shell) was discovered in 2021, allowing attackers to remotely execute code on vulnerable systems.

related terms: Vulnerability, Zero-Day, Patch

Logic Bomb

definition: A piece of code intentionally inserted into a software system that will set off a malicious function when specified conditions are met.

explanation: It is like a time bomb that is set to go off at a specific time or under certain conditions.

real-world examples: A disgruntled employee might plant a logic bomb in a company's system that triggers when their employment is terminated.

related terms: Malware, Time Bomb, Trojan Horse

Logic Error

definition: A mistake in a program's source code resulting in wrong or unexpected behavior.

explanation: It is like a typo in a recipe that causes the dish to taste bad.

real-world examples: A calculation error in a banking application that results in incorrect account balances.

related terms: Software Bug, Debugging, Software Testing

Logical Security

definition: The use of software and data to protect computer systems and networks from unauthorized access.

explanation: It is like the digital locks, passwords, and alarms that protect a building.

real-world examples: Firewalls, intrusion detection systems, antivirus software, and access controls are examples of logical security measures.

related terms: Cybersecurity, Physical Security, Network Security

MAC Address Filtering

definition: A security access control method where the 48-bit address assigned to each network card is utilized to establish access to the network.

explanation: Like a VIP list for a party, only devices with specific MAC addresses are allowed on the network.

real-world examples: Used in Wi-Fi networks to restrict access to authorized devices.

related terms: Network Security, Access Control, Whitelisting

MAC Address Spoofing

definition: An attack technique that involves changing a device's Media Access Control (MAC) address.

explanation: It is like changing your name tag at a party to impersonate someone else.

real-world examples: Attackers use MAC address spoofing to bypass filters, gain unauthorized access to networks, or impersonate other devices.

related terms: Spoofing, Identity Theft, Network Security

MAC Flooding

definition: An attack that overwhelms a network switch's MAC address table, causing it to flood all ports with traffic.

explanation: It is like a stampede of people rushing through a single door, causing chaos and disruption.

real-world examples: Attackers use MAC flooding to disrupt network traffic, eavesdrop on communications, or launch other attacks.

related terms: Denial of Service (DoS), Switch, Network Security

Machine Learning (ML)

definition: A kind of artificial intelligence (AI) that lets software applications to be more accurate in predicting outcomes without being explicitly programmed to do so.

explanation: It is like teaching a computer to learn from experience, so it can improve its performance over time.

real-world examples: Used in spam filters, fraud detection systems, and recommendation engines.

related terms: Artificial Intelligence (AI), Deep Learning, Neural Networks

Machine Learning Security

definition: The practice of protecting machine learning models and systems from attacks and vulnerabilities.

explanation: It is like putting a security system in place to protect a self-learning robot from being tampered with or misled.

real-world examples: Defending against adversarial machine learning attacks, ensuring the integrity of training data, and protecting against model theft.

related terms: Adversarial Machine Learning, Cybersecurity, Artificial Intelligence

Macro Virus

definition: A computer virus that is written in a macro language and embedded in a document or application.

explanation: It is like a booby trap hidden inside a harmless-looking file.

real-world examples: Macro viruses often spread through email attachments and can infect your computer when you open the document.

related terms: Virus, Malware, Microsoft Office

Magecart Attacks

definition: Cyberattacks that target online shopping cart systems to steal customer payment information.

explanation: Imagine a thief lurking in the checkout line of an online store, secretly stealing your credit card details as you enter them.

real-world examples: Magecart groups have targeted major retailers like British Airways and Ticketmaster, compromising millions of customer records.

related terms: E-commerce Fraud, Web Skimming, Credit Card Theft

Malicious Attachment

definition: An email attachment that contains malware or other malicious code.

explanation: It is like a poisoned apple - it looks tempting, but it is actually harmful.

real-world examples: Attackers often use social engineering techniques to trick users into opening malicious attachments.

related terms: Phishing, Malware, Email Security

Malicious Code

definition: Any code that is created to harm a computer system or network.

explanation: It is like a computer virus - it can infect your system and cause all sorts of problems.

real-world examples: Malware, viruses, worms, Trojans, and ransomware.

related terms: Malware, Cyber Attack, Vulnerability

Malicious Domain

definition: A domain name that is used for malicious purposes, such as phishing, malware distribution, or command-and-control (C2) activity.

explanation: It is like a fake address used by a criminal to lure victims into a trap.

real-world examples: Often used in phishing scams to trick users into thinking they are visiting a legitimate website.

related terms: Phishing, Malware, Domain Name System (DNS)

Malicious Insider

definition: A current or former employee, contractor, or business partner who has or had authorized access to an organization's network, system, or data and intentionally exceeded or misused that access, negatively affecting the confidentiality, integrity, or availability of the organization's information or information systems.

explanation: It is like a spy who has infiltrated your organization and is working from the inside to steal information or cause damage.

real-world examples: Employees who steal trade secrets, sabotage systems, or leak confidential data.

related terms: Insider Threat, Data Breach, Espionage

Malicious URL

definition: A web address that leads to a malicious website or downloads malware.

explanation: It is like a trapdoor that leads to a dangerous place.

real-world examples: Often used in phishing scams to trick users into visiting fake websites or downloading malware.

related terms: Phishing, Malware, URL Filtering

Malvertising

definition: The use of online advertising to distribute malware or redirect users to malicious websites.

explanation: It is like a poisoned ad in a magazine – you flip through the pages, and suddenly you're infected with a virus.

real-world examples: Attackers compromise legitimate advertising networks to display malicious ads on popular websites.

related terms: Malware, Drive-by Download, Online Advertising

Malware

definition: Malicious software designed to harm a computer system or network.

explanation: It is like a disease that infects your computer, causing it to malfunction or steal your data.

real-world examples: Viruses, worms, Trojans, ransomware, spyware, and adware.

related terms: Cybersecurity, Virus, Cyber Attack

Malware Analysis

definition: The process of studying malware to understand its behavior, capabilities, and origin.

explanation: It is like a doctor examining a virus under a microscope to figure out how it works and how to treat it.

real-world examples: Used to develop antivirus software, identify the source of attacks, and gather intelligence on cyber threats.

related terms: Reverse Engineering, Sandboxing, Threat Intelligence

Malware as a Service (MaaS)

definition: A business model where cybercriminals offer malware and related services to other criminals for a fee.

explanation: It is like a subscription service for cybercrime, where you can pay to rent a botnet or purchase a ransomware kit.

real-world examples: Allows less-skilled attackers to launch sophisticated cyber attacks without having to develop their own malware or infrastructure.

related terms: Cybercrime, Ransomware, Botnet

Malware Removal

definition: The process of eliminating malicious software (malware) from a computer system or network.

explanation: Imagine it as disinfecting a wound to remove harmful bacteria. Malware removal tools identify and eliminate various types of malwares, such as viruses, worms, Trojans, and ransomware.

real-world examples: Using antivirus or anti-malware software to scan and remove malicious code from a computer.

related terms: Antivirus Software, Malware, Remediation

Malware Sandbox

definition: An isolated environment where suspicious files or code can be executed and analyzed without harming the actual system.

explanation: Think of it as a quarantine zone for potentially dangerous software, where it can be safely observed and studied.

real-world examples: Security researchers use sandboxes to analyze new malware strains and understand their behavior.

related terms: Malware Analysis, Virtual Machine, Emulation

Malware Signature

definition: A unique pattern of code or data that identifies a specific piece of malware.

explanation: It is like a fingerprint for malware that security software can use to detect and identify it.

real-world examples: Antivirus software uses signature databases to identify known malware threats.

related terms: Malware Detection, Antivirus, Signature-Based Detection

Managed Detection and Response (MDR)

definition: A service that provides continuous threat monitoring and response through a team of experts.

explanation: It's like having a team of security experts watching over your network 24/7.

real-world examples: Outsourcing threat detection and response to a specialized provider to enhance your organization's security.

related terms: Security Operations Center (SOC), Incident Response, Threat Monitoring, Cybersecurity Services

Man-in-the-Browser (MitB) Attack

definition: An attack where malware infects a web browser to intercept and modify traffic between the user and a website.

explanation: Imagine a malicious actor eavesdropping on your conversations with your bank, potentially altering your transactions.

real-world examples: Attackers can steal login credentials, modify online banking transactions, or inject malicious code into web pages.

related terms: Man-in-the-Middle Attack, Browser Exploit, Malware

Man-in-the-Cloud Attack

definition: An attack that targets cloud services and infrastructure to gain unauthorized access to data or resources.

explanation: It is like a hacker breaking into a shared apartment building and accessing multiple tenants' belongings.

real-world examples: Attackers can exploit vulnerabilities in cloud platforms, steal credentials, or compromise cloud-based applications.

related terms: Cloud Security, Data Breach, Cloud Computing Attack

Man-in-the-Middle (MitM) Attack

definition: An attack where the attacker secretly relays and possibly alters the communication between two parties who believe they are directly communicating with each other.

explanation: Imagine a postman who secretly opens your mail, reads it, possibly changes it, and then reseals it before delivering it to the intended recipient.

real-world examples: Attackers can use MitM attacks to eavesdrop on conversations, steal sensitive information, or inject malicious code.

related terms: Eavesdropping, Packet Sniffing, ARP Spoofing

Masquerade Attack

definition: An attack where an attacker impersonates a trusted entity to gain unauthorized access or privileges.

explanation: It is like wearing a disguise to sneak into a party you weren't invited to.

real-world examples: Attackers might use stolen credentials, forged IP addresses, or spoofed emails to impersonate legitimate users or systems.

related terms: Spoofing, Identity Theft, Social Engineering

Master Boot Record (MBR)

definition: The first sector of a data storage device that contains information about the device's partitions and how to boot the operating system.

explanation: It is like the table of contents for your computer's hard drive, telling it where to find the operating system and how to start it up.

real-world examples: Malware can infect the MBR to prevent the operating system from booting or to redirect the boot process to a malicious operating system.

related terms: Boot Sector, Bootloader, Rootkit

Master Service Agreement (MSA)

definition: A contract between a service provider and a customer outlining the terms and conditions of their relationship.

explanation: It is like a legal agreement between a landlord and a tenant, defining the rights and responsibilities of both parties.

real-world examples: The MSA typically covers aspects like service levels, pricing, and dispute resolution.

related terms: Service Level Agreement (SLA), Contract, Outsourcing

Mean Time to Detect (MTTD)

definition: The average time it takes to detect a security incident.

explanation: It is like the time it takes for a smoke detector to go off after a fire starts. The faster It is detected, the quicker you can respond.

real-world examples: A security team might measure their MTTD for malware infections or phishing attacks.

related terms: Security Metrics, Threat Detection, Incident Response

Mean Time to Respond (MTTR)

definition: The average time it takes to respond to and resolve a security incident.

explanation: It is like the time it takes for firefighters to arrive at a fire and extinguish it. The faster they respond; the less damage is done.

real-world examples: A company might track their MTTR for different types of incidents, such as data breaches or system outages.

related terms: Security Metrics, Incident Response, Service Level Agreement (SLA)

Meet-in-the-Middle Attack

definition: A cryptographic attack that combines a known plaintext attack with a brute-force attack to reduce the complexity of breaking a cryptographic algorithm.

explanation: Imagine you have a combination lock with a 10-digit code. Instead of trying all 10 billion combinations, you might try the first 5 digits and the last 5 digits separately, significantly reducing the number of combinations to try.

real-world examples: Used to attack encryption algorithms like Double DES and Triple DES.

related terms: Cryptanalysis, Encryption, Brute Force Attack

Memory Corruption

definition: A vulnerability that occurs when a program modifies or accesses memory in an unintended way.

explanation: It is like accidentally spilling coffee on your notes – the information becomes garbled and unusable. In this case, the "mess" is an opportunity for an attacker to execute malicious code.

real-world examples: Buffer overflows, buffer underflows, and use-after-free vulnerabilities can all lead to memory corruption.

related terms: Vulnerability, Exploit, Software Bug

Memory Dump

definition: A snapshot of the contents of a computer's memory at a specific point in time.

explanation: It is like taking a picture of your computer's "brain" to see what it was thinking at that moment.

real-world examples: Used for debugging, troubleshooting, and forensic analysis.

related terms: Memory Forensics, Crash Dump, Debugging

Memory Forensics

definition: The analysis of a computer's memory to investigate security incidents or recover evidence.

explanation: It is like examining the brain of a computer to see what it was thinking at the time of a crime.

real-world examples: Used to recover passwords, encryption keys, and other sensitive data from a computer's memory.

related terms: Digital Forensics, Memory Dump, Volatility Framework

Memory Leak

definition: A bug in a program that causes it to consume more and more memory over time, eventually leading to a crash.

explanation: It is like a leaky faucet that slowly fills up a bathtub until it overflows.

real-world examples: Memory leaks can cause performance problems, system instability, and even security vulnerabilities.

related terms: Software Bug, Resource Leak, Debugging

Message Authentication Code (MAC)

definition: A short piece of information used to authenticate a message and ensure its integrity.

explanation: It is like a checksum for a message, but it is created using a secret key to ensure that only the intended recipient can verify its authenticity.

real-world examples: Used to protect the integrity and authenticity of messages in transit, such as financial transactions or sensitive communications.

related terms: Cryptography, Hash Function, Authentication

Metadata

definition: Data that provides information about other data.

explanation: It is like the label on a food package – it tells you about the contents, but it is not the food itself.

real-world examples: The metadata of a file might include its size, creation date, and author. In photos, it might include location, camera settings, and time taken.

related terms: Data, Information, File System

Metamorphic Malware

definition: Malicious software that changes its code each time it replicates, making it harder to detect.

explanation: Think of it like a chameleon that constantly changes its colors to avoid detection. This type of malware is particularly tricky for antivirus software to detect, as it doesn't have a fixed signature.

real-world examples: Examples include the Zmist and Virlock malware families.

related terms: Malware, Polymorphic Malware, Antivirus Evasion

Metasploit

definition: A penetration testing framework that provides tools for exploiting vulnerabilities in systems and networks.

explanation: It is like a toolbox for ethical hackers, containing a wide range of tools for testing security defenses.

real-world examples: Used by security professionals to simulate attacks, identify vulnerabilities, and develop exploits.

related terms: Penetration Testing, Vulnerability Assessment, Exploit Framework

Microsegmentation

definition: Dividing a network into smaller parts to improve security.

explanation: It's like creating small, secure zones within your network to prevent attackers from moving freely.

real-world examples: Isolating sensitive data and applications in a data center to protect them from attacks.

related terms: Network Segmentation, Zero Trust, Software-Defined Networking (SDN), Access Control

MITRE ATT&CK

definition: A globally accessible knowledge base of adversary tactics and techniques that are based on real-world observations.

explanation: Think of it as an encyclopedia of cyber attacker behavior, outlining the steps they take to compromise systems and what they do once they're in.

real-world examples: Used by security professionals to better understand attacker behavior and develop more effective defenses.

related terms: Threat Intelligence, Cyber Attack, Threat Modeling

Mobile Application Management (MAM)

definition: Software that allows IT administrators to manage and secure mobile applications on corporate or personally owned devices.

explanation: It is like a security guard for your apps, ensuring that only authorized apps are installed, and that sensitive data is protected.

real-world examples: Used to control which apps can be installed on a device, manage app configurations, and enforce security policies.

related terms: Mobile Device Management (MDM), Application Security, BYOD

Mobile Application Security

definition: The practice of securing mobile applications from threats and vulnerabilities.

explanation: It is like putting a lock on your mobile app to protect it from being, and code signing.

real-world examples: Application Security, Mobile Security, Mobile Malware

related terms: Mobile Security, Application Security,

Mobile Device Exploits

definition: Attacks that target vulnerabilities in mobile devices in order to gain unauthorized access or control.

explanation: It's like picking the lock on a smartphone.

real-world examples: Exploiting vulnerabilities in mobile operating systems to install spyware.

related terms: Mobile Security, Cybersecurity, Exploits

Mobile Device Management (MDM)

definition: Software that allows IT administrators to manage and secure mobile devices utilized in the workplace.

explanation: It is like a set of parental controls for company-issued devices, allowing IT to enforce security policies and remotely manage devices.

real-world examples: Enforcing password policies, installing updates, and remotely wiping lost or stolen devices.

related terms: Mobile Security, BYOD, Endpoint Security

Mobile Malware

definition: Malicious software designed to target mobile devices, such as smartphones and tablets.

explanation: It is like a virus that infects your phone, stealing data, sending premium SMS messages, or even spying on you.

real-world examples: Android malware, iOS malware, spyware, and ransomware.

related terms: Malware, Mobile Security, Cybersecurity

Mobile Security

definition: The practice of protecting mobile devices from unauthorized access, data loss, and other threats.

explanation: It is like putting a security system on your phone to keep it safe from thieves and hackers.

real-world examples: Using strong passwords, installing security apps, and keeping software up to date.

related terms: Cybersecurity, Mobile Device Management (MDM), Mobile Threat Defense (MTD)

Mobile Threat Defense (MTD)

definition: A security solution that protects mobile devices from threats such as malware, phishing attacks, and network attacks.

explanation: It is like a bodyguard for your mobile device, constantly watching for danger and taking action to protect you.

real-world examples: MTD solutions use a variety of techniques, such as antivirus scanning, threat intelligence, and behavioral analysis.

related terms: Mobile Security, Endpoint Security, Malware Protection

Multi-factor Authentication (MFA)

definition: An authentication method that requires the user to provide two or more verification factors to gain access to a system or application.

explanation: It is like having multiple locks on your front door - it makes it much harder for someone to break in.

real-world examples: Using a password and a fingerprint scan, or a password and a one-time code sent to your phone.

related terms: Authentication, Two-Factor Authentication (2FA), Security

Multipartite Virus

definition: A type of computer virus that can infect both boot sectors and executable files.

explanation: It is like a virus that can spread through multiple channels, making it more difficult to contain.

real-world examples: Can infect both the boot sector of a hard drive and executable files, making it difficult to remove.

related terms: Virus, Malware, Boot Sector Virus

Multi-Party Computation (MPC)

definition: A cryptographic protocol that allows multiple parties to jointly compute a function over their inputs while keeping those inputs private.

explanation: Imagine a group of friends trying to calculate their average salary without revealing their individual incomes. MPC makes this possible.

real-world examples: Used in secure voting systems, private auctions, and other applications where privacy is important.

related terms: Cryptography, Privacy, Secure Computation

Multi-Tenancy

definition: A software architecture in which a single instance of a software application serves multiple customers.

explanation: It is like a shared apartment building - multiple tenants live in the same building, but each has their own private space.

real-world examples: Cloud computing providers often use multi-tenancy to deliver their services to multiple customers.

related terms: Cloud Computing, Software as a Service (SaaS), Scalability

Multi-Vector Attack

definition: An attack that uses multiple attack vectors to exploit vulnerabilities in a system or network.

explanation: It's like a burglar trying to break into a house through multiple entry points - windows, doors, chimneys, etc.

real-world examples: An attacker might use a combination of phishing emails, malware, and social engineering to compromise a target.

related terms: Attack Vector, Threat, Vulnerability

Mutual Authentication

definition: An authentication process where both parties verify the identity of the other.

explanation: It is like two spies exchanging secret codes to confirm they are who they say they are.

real-world examples: Used in secure communication protocols like TLS to prevent man-in-the-middle attacks.

related terms: Authentication, SSL/TLS, Man-in-the-Middle Attack

N

National Institute of Standards and Technology (NIST)

definition: A U.S. government agency that develops and promotes measurement, standards, and technology to enhance productivity, facilitate trade, and improve the quality of life.

explanation: They are like the referees of the technology world, setting the standards and guidelines for everything from cybersecurity to weights and measures.

real-world examples: NIST develops and maintains the Cybersecurity Framework, a set of guidelines for managing cybersecurity risk.

related terms: Cybersecurity, Standards, U.S. Government

National Vulnerability Database (NVD)

definition: The U.S. government repository of standards-based vulnerability management data represented using the Security Content Automation Protocol (SCAP).

explanation: It is like a public library of security vulnerabilities, providing detailed information about known weaknesses in software and hardware.

real-world examples: Security professionals use the NVD to identify and prioritize vulnerabilities that need to be patched.

related terms: Vulnerability, CVE, Security Advisory

Nation-State Actor

definition: An individual or group that is supported or sponsored by a nation-state to conduct cyber operations.

explanation: These are the government-backed hackers, often with significant resources and sophisticated tools at their disposal.

real-world examples: Nation-state actors are responsible for many high-profile cyberattacks, including espionage, sabotage, and election interference.

related terms: Cyber Espionage, Cyber Warfare, Advanced Persistent Threat (APT)

Nested Virtualization

definition: A type of virtualization where a virtual machine (VM) runs inside another virtual machine.

explanation: It is like a Russian nesting doll – one VM is nested inside another, which is nested inside another, and so on.

real-world examples: Used for testing and development purposes, as well as for creating isolated environments for running sensitive applications.

related terms: Virtualization, Virtual Machine (VM), Hypervisor

Network Access Control (NAC)

definition: A security approach that controls which devices can access a network and what they can do once connected.

explanation: Imagine a bouncer at a club, not only checking IDs but also assigning different access levels to guests (VIP, dance floor only, etc.).

real-world examples: Employees' laptops are granted access to specific company resources, while guest devices are restricted to internet access only.

related terms: Authentication, Authorization, 802.1X

Network Address Hijacking

definition: An attack where an attacker takes control of a legitimate IP address to intercept or manipulate network traffic.

explanation: Like a hijacker taking over a plane, the attacker gains control of the IP address and can redirect its traffic.

real-world examples: Attackers can intercept sensitive information, inject malicious code into traffic, or launch denial-of-service attacks.

related terms: IP Spoofing, Man-in-the-Middle Attack, ARP Spoofing

Network Address Translation (NAT)

definition: A method of remapping one IP address space into another by altering network address information in the IP header of packets while they are in transit across a traffic routing device.

explanation: Think of it like a mailroom that changes the recipient's address on envelopes before forwarding them. It allows multiple devices on a private network to share a single public IP address.

real-world examples: Used in home routers to connect multiple devices to the internet using a single IP address provided by the ISP.

related terms: IP Address, Port Forwarding, Firewall

Network Anomaly Detection

definition: A security technique that identifies unusual or suspicious patterns in network traffic.

explanation: It is like a security camera that alerts you when someone is acting suspiciously in your home.

real-world examples: Detecting unusual traffic spikes, unusual protocols, or communication with known malicious IP addresses.

related terms: Intrusion Detection System (IDS), Machine Learning, Network Security Monitoring

Network Architecture

definition: The design and structure of a computer network, including the physical and logical layout of its elements.

explanation: It's like the blueprint for building a complex highway system, showing how different roads (networks) connect and interact.

real-world examples: Designing a company's network to include routers, switches, firewalls, and wireless access points to support business operations.

related terms: Network Design, Network Topology, IT Infrastructure

Network Behavior Analysis (NBA)

definition: A security technique that analyzes network traffic patterns to identify and respond to potential threats.

explanation: It is like a detective analyzing surveillance footage to identify suspicious behavior.

real-world examples: Monitoring network traffic for signs of malware, botnets, or other malicious activity.

related terms: Network Anomaly Detection, Threat Intelligence, Security Information and Event Management (SIEM)

Network Firewall

definition: A security system monitoring and controlling incoming and outgoing network traffic based on predetermined security rules.

explanation: It's like a security guard at the entrance of a building, checking everyone's ID and only allowing authorized personnel to enter.

real-world examples: Blocking unauthorized access attempts, filtering malicious traffic, and protecting against network-based attacks.

related terms: Firewall, Network Security, Intrusion Prevention System (IPS)

Network Flow Analysis

definition: The process of collecting and analyzing network traffic data to gain insights into network usage and security.

explanation: It's like analyzing traffic patterns on a highway to understand where cars are coming from, where they're going, and how fast they're traveling.

real-world examples: Used to identify network bottlenecks, detect anomalies, and troubleshoot performance problems.

related terms: Network Traffic Analysis, Security Information and Event Management (SIEM), NetFlow

Network Forensics

definition: The process of investigating and analyzing network traffic to gather evidence of a crime or security incident.

explanation: It's like a detective analyzing traffic camera footage to reconstruct a crime scene.

real-world examples: Used to trace the source of an attack, identify compromised systems, and gather evidence for legal proceedings.

related terms: Digital Forensics, Incident Response, Security Investigation

Network Hardening

definition: The process of securing a network by reducing its vulnerabilities and attack surface.

explanation: It's like reinforcing the walls of a castle to make it more difficult for attackers to breach.

real-world examples: Disabling unnecessary services, applying security patches, and configuring firewalls are all examples of network hardening techniques.

related terms: Security Hardening, Vulnerability Management, Patch Management

Network Intrusion

definition: Unauthorized access to a computer network.

explanation: It's like breaking into a secure building to steal information.

real-world examples: Hackers gaining access to a corporate network to steal data.

related terms: Cybersecurity, Network Security, Intrusion Detection

Network Intrusion Detection System (NIDS)

definition: A system monitoring network traffic for suspicious activity and alerting security personnel to potential threats.

explanation: It's like a burglar alarm that goes off when it detects someone trying to break into your house.

real-world examples: Sniffing network traffic for signatures of known attacks, analyzing traffic patterns for anomalies, and alerting administrators to potential security breaches.

related terms: Intrusion Detection System (IDS), Network Security, Threat Detection

Network Intrusion Prevention System (NIPS)

definition: A system monitoring network traffic for suspicious activity and actively blocks or mitigates potential threats.

explanation: It's like a security guard who not only monitors intruders but also tackles them if they try to enter the building.

real-world examples: NIPS can block malicious traffic, quarantine infected files, and take other actions to prevent attacks from succeeding.

related terms: Intrusion Prevention System (IPS), Firewall, Network Security

Network Mapper (Nmap)

definition: A network scanning tool used to discover hosts and services on a computer network.

explanation: It's like a radar system that scans the network for devices and services, creating a map of the network topology.

real-world examples: Used by network administrators for network inventory and security auditing, as well as by attackers for reconnaissance.

related terms: Port Scanning, Network Scanning, Vulnerability Scanning

Network Perimeter

definition: The boundary between an organization's internal network and the external world, such as the internet.

explanation: It's like the walls of a castle, separating the protected interior from the outside world.

real-world examples: Firewalls, intrusion detection and prevention systems, and demilitarized zones (DMZs) are all used to secure the network perimeter.

related terms: Network Security, Perimeter Security, DMZ

Network Reconnaissance

definition: The process of gathering information about a target network, often as a precursor to a cyber attack.

explanation: It's like a spy chasing a target before launching an operation, looking for weaknesses and vulnerabilities.

real-world examples: Port scanning, vulnerability scanning, and social engineering are all examples of network reconnaissance techniques.

related terms: Penetration Testing, Ethical Hacking, Threat Intelligence

Network Security

definition: The practice of protecting computer networks from unauthorized access, misuse, modification, or denial of service.

explanation: It's like a security system for your network, protecting it from intruders and malicious activity.

real-world examples: Firewalls, intrusion detection and prevention systems, antivirus software, and encryption are all used to ensure network security.

related terms: Cybersecurity, Information Security, Network Hardening

Network Security Monitoring

definition: The process of continuously monitoring network traffic and systems for signs of security threats.

explanation: It's like having a security camera that watches your network 24/7, alerting you to any suspicious activity.

real-world examples: Used to detect and respond to cyberattacks, identify vulnerabilities, and troubleshoot network problems.

related terms: Security Information and Event Management (SIEM), Intrusion Detection System (IDS), Threat Detection

Network Segmentation

definition: The act of dividing a computer network into smaller subnetworks, each with its own security controls.

explanation: It's like dividing a city into different neighborhoods, each with its own security measures.

real-world examples: Used to isolate critical systems, limit the spread of malware, and enforce security policies.

related terms: Network Security, Firewall, Access Control

Network Spoofing

definition: Creating a fake network to trick users into connecting and revealing sensitive information.

explanation: It's like setting up a fake toll booth to collect money from unsuspecting drivers.

real-world examples: Setting up a fake Wi-Fi hotspot to intercept user data.

related terms: Social Engineering, Cybersecurity, Network Security

Network Tap

definition: A hardware device that allows you to monitor network traffic without affecting its flow.

explanation: It's like a traffic camera that records everything that passes by without interfering with the traffic itself.

real-world examples: Used for network monitoring, intrusion detection, and security analysis.

related terms: Network Monitoring, Packet Sniffing, Security Monitoring

Network Time Protocol (NTP)

definition: A networking protocol for clock synchronization between computer systems over packet-switched, variable-latency data networks.

explanation: It's like a conductor synchronizing the instruments in an orchestra. NTP helps computers agree on the time, which is important for logging events, security protocols, and time-sensitive transactions.

real-world examples: Used to synchronize clocks on servers, workstations, and network devices.

related terms: Clock Synchronization, Time Server, Network Protocol

NFC Attack

definition: An attack that targets near-field communication (NFC) technology, which is used for contactless payments and data transfer.

explanation: It's like a digital pickpocket stealing information from your contactless card or phone.

real-world examples: Attackers using NFC readers to steal credit card information from contactless cards or intercepting data transmitted between NFC-enabled devices.

related terms: Contactless Payment, RFID Skimming, Data Theft

NIST Cybersecurity Framework (CSF)

definition: A voluntary framework consisting of standards, guidelines, and best practices in order to manage cybersecurity risk.

explanation: It's like a toolbox for building and maintaining a strong cybersecurity posture.

real-world examples: Organizations using the NIST CSF to develop and improve their cybersecurity programs.

related terms: Cybersecurity, Risk Management, Best Practices

NIST SP 800-53

definition: A catalog of security and privacy controls for federal information systems and organizations.

explanation: It's like a comprehensive guide to implementing security controls in government systems.

real-world examples: Federal agencies adopting NIST SP 800-53 controls to secure their information systems.

related terms: Cybersecurity, Information Security, Regulatory Compliance

Nonce

definition: A random or semi-random number that is utilized only once in a cryptographic communication.

explanation: It's like a unique key that is used once and then discarded.

real-world examples: Nonces used in cryptographic protocols to prevent replay attacks.

related terms: Cryptography, Encryption, Security Token

Non-Disclosure Agreement (NDA)

definition: A legal contract that prohibits the disclosure of confidential information.

explanation: It's like pinky swear but with legal consequences. NDAs are used to protect trade secrets, business plans, and other sensitive information.

real-world examples: Employees sign NDAs to protect their employer's confidential information.

related terms: Confidentiality Agreement, Trade Secret, Intellectual Property

Non-repudiation

definition: The assurance that someone cannot deny the validity of something.

explanation: Imagine a digital signature on an email that proves you sent it and can't later deny doing so.

real-world examples: Digital signatures and audit logs are used to provide non-repudiation.

related terms: Authentication, Digital Signature, Audit Log

NoSQL Injection

definition: A type of injection attack targeting NoSQL databases.

explanation: It's like SQL injection, but for NoSQL databases. Attackers inject malicious code into queries to access or modify data.

real-world examples: MongoDB and CouchDB are vulnerable to NoSQL injection attacks.

related terms: Injection Attack, NoSQL Database, Web Application Security

Obfuscation

definition: The process of making something obscure or unclear. In cybersecurity, it's often used to hide code or data from prying eyes.

explanation: It's like camouflaging an object to make it blend in with its surroundings.

real-world examples: Obfuscation can be used to protect intellectual property or to make it harder for attackers to understand how malware works.

related terms: Obfuscated Code, Code Obfuscation, Malware

Object-Level Security

definition: A security model that applies permissions to individual objects within a system, such as files, folders, or database records.

explanation: Think of it like putting locks on individual drawers in a filing cabinet, rather than just locking the whole cabinet.

real-world examples: Used to control access to sensitive data and ensure that only authorized users can view or modify it.

related terms: Access Control, Authorization, Data Security

Offline Brute Force Attack

definition: A type of brute force attack where the attacker has a copy of the encrypted data and can try to crack it offline.

explanation: It's like trying to guess the combination to a lock by taking it home and trying different combinations in private.

real-world examples: Attackers might steal a password hash database and then use powerful computers to try to crack the passwords offline.

related terms: Brute Force Attack, Password Cracking, Hash

On-Demand Self-Service

definition: A cloud computing characteristic that allows users to provision computing resources (such as server time or network storage) automatically without requiring human interaction with each service provider.

explanation: It's like a vending machine for computing resources – users can select what they need and get it instantly, without having to wait for someone to fulfill their request.

real-world examples: Cloud providers like AWS and Azure offer on-demand self-service for virtual machines, storage, and other resources.

related terms: Cloud Computing, Elasticity, Scalability

One-Time Password (OTP)

definition: A password that is valid for only one login session or transaction.

explanation: It's like a single-use ticket to a concert – once you've used it, it's no longer valid.

real-world examples: Used for two-factor authentication (2FA) and other security mechanisms to add an extra layer of protection.

related terms: Authentication, Two-Factor Authentication (2FA), Security Token

Online Brute Force Attack

definition: A type of brute force attack where the attacker tries to guess a password or encryption key by repeatedly submitting guesses through a login interface.

explanation: It's like trying to guess a password by typing it into a login screen repeatedly.

real-world examples: Attackers often use automated tools to launch online brute force attacks, trying thousands of passwords per second.

related terms: Brute Force Attack, Password Cracking, Account Lockout

Open Port

definition: A TCP or UDP port number that is open and accepting connections from remote hosts.

explanation: It's like an open door in your computer's firewall, allowing traffic to enter and exit.

real-world examples: Open ports can be used for legitimate purposes, such as web browsing (port 80) or email (port 25) but can also be exploited by attackers.

related terms: Firewall, Port Scanning, Network Security

Open Proxy

definition: A proxy server that is accessible to anyone on the internet.

explanation: It's like a public phone booth – anyone can use it to make calls.

real-world examples: Often used to anonymize web traffic or bypass content filters.

related terms: Proxy Server, Anonymity, Web Proxy

Open Redirect

definition: A vulnerability in a web application that allows an attacker to redirect users to a malicious website.

explanation: It's like a detour sign that leads you down a dangerous road.

real-world examples: Attackers can use open redirects to trick users into visiting phishing websites or downloading malware.

related terms: Web Application Security, Vulnerability, Phishing

Open-Source Intelligence (OSINT)

definition: Information that is collected from publicly available sources.

explanation: It's like conducting research using newspapers, magazines, and websites.

real-world examples: Social media posts, news articles, and government reports are all examples of OSINT.

related terms: Threat Intelligence, Information Gathering, Reconnaissance

Open-Source Security

definition: A philosophy and practice of developing and using security software that is freely available and open to public scrutiny.

explanation: It's like a community garden – everyone can contribute to its growth and maintenance.

real-world examples: Open-source security tools like Wireshark, Snort, and OpenVAS are widely used by security professionals.

related terms: Open-Source Software, Cybersecurity, Community-Driven Development

Open-Source Security Testing Methodology Manual (OSSTMM)

definition: A peer-reviewed methodology for security testing and analysis.

explanation: It's like a cookbook for security testing, providing detailed instructions on how to conduct different types of assessments.

real-world examples: Used by security professionals to plan and execute security tests.

related terms: Security Testing, Penetration Testing, Vulnerability Assessment

Open-Source Vulnerability Database (OSVDB)

definition: A database of vulnerabilities in open-source software.

explanation: It's like a catalog of known defects in open-source software, similar to the National Vulnerability Database (NVD) for commercial software.

real-world examples: Used by security professionals to identify and prioritize vulnerabilities that need to be patched.

related terms: Vulnerability, Open-Source Software, Security Advisory

Open Systems Interconnection (OSI) Model

definition: A conceptual framework that standardizes the functions of a telecommunication or computing system into seven different layers.

explanation: It's like a blueprint for how computer networks communicate with each other, with each layer responsible for a specific set of tasks.

real-world examples: The seven layers are: Physical, Data Link, Network, Transport, Session, Presentation, and Application.

related terms: Network Protocol, Networking, TCP/IP

Open Web Application Security Project (OWASP)

definition: A non-profit foundation dedicated to improving software security.

explanation: A global community of security experts who create free resources (like the OWASP Top 10) to help organizations build more secure applications.

real-world examples: The OWASP Top 10 list of most critical web application security risks is a key resource for developers and security professionals.

related terms: Web Application Security, Vulnerability, Security Standards

Operational Audit

definition: A comprehensive review of the operations of a business to evaluate efficiency, effectiveness, and economy.

explanation: It's like a performance review for an organization's processes and operations.

real-world examples: Auditing the efficiency of supply chain processes.

related terms: Internal Audit, Compliance, Risk Management

Operational Resilience

definition: The ability of an organization to continue delivering critical operations through disruption.

explanation: It's like having a backup plan to ensure business continuity despite unexpected events.

real-world examples: Companies developing business continuity plans to handle natural disasters and cyberattacks.

related terms: Business Continuity Planning (BCP), Risk Management, Incident Response

Operational Security (OPSEC)

definition: The process of identifying and protecting critical information to prevent it from falling into the wrong hands.

explanation: Think of it like a game of hide-and-seek with your secrets - you want to keep your sensitive information hidden from prying eyes.

real-world examples: Military operations use OPSEC to conceal troop movements, and businesses use it to protect trade secrets.

related terms: Information Security, Risk Management, Counterintelligence

Operational Technology (OT)

definition: Hardware and software systems monitoring and controlling physical devices, processes, and events in the enterprise.

explanation: It's the technology that makes things happen in the real world, like manufacturing equipment, power grids, and transportation systems.

real-world examples: SCADA systems, industrial control systems (ICS), and building automation systems.

related terms: Industrial Control System (ICS), Critical Infrastructure, SCADA

Outbound Firewall

definition: A network security device monitoring and controlling outgoing network traffic.

explanation: Think of it as a one-way security gate - it lets traffic out but carefully checks everything that tries to leave.

real-world examples: Used to prevent data exfiltration and block unauthorized communication from internal systems to the internet.

related terms: Firewall, Network Security, Data Loss Prevention (DLP)

Out-of-Band (OOB) Authentication

definition: An authentication method that uses a separate communication channel from the primary channel to verify a user's identity.

explanation: It's like a secret knock that you use to confirm someone's identity before opening the door.

real-world examples: Sending a one-time code to a user's phone during login or using a hardware token to generate a verification code.

related terms: Two-Factor Authentication (2FA), Multi-Factor Authentication (MFA), Authentication

Out-of-Band Attack

definition: An attack that exploits a vulnerability in a system or protocol that is not directly related to the primary communication channel.

explanation: It's like a burglar finding an unlocked window on the side of a house while everyone is focused on the front door.

real-world examples: DNS tunneling, VoIP (Voice over IP) attacks, and attacks that exploit vulnerabilities in network protocols.

related terms: Vulnerability, Exploit, Attack Vector

Overflow Attack

definition: An attack that exploits a buffer overflow or integer overflow vulnerability to execute malicious code or crash a system.

explanation: It's like overfilling a glass of water – the excess spills out and can cause a mess. In this case, the "mess" is an opportunity for an attacker to take control of the system.

real-world examples: Buffer overflow attacks are a common way for attackers to gain unauthorized access to systems and execute arbitrary code.

related terms: Buffer Overflow, Buffer Underflow, Integer Overflow

Over-the-Air (OTA) Update

definition: A method of distributing software updates to mobile devices wirelessly.

explanation: It's like updating your phone's apps without having to plug it into a computer.

real-world examples: OTA updates are used to patch vulnerabilities, add new features, and improve performance.

related terms: Software Update, Mobile Device Management (MDM), Patch Management

Over-the-Shoulder Attack

definition: A type of social engineering attack where an attacker observes a victim's screen or keyboard input to steal sensitive information.

explanation: It's like someone looking over your shoulder as you enter your PIN at an ATM.

real-world examples: Attackers might try to steal passwords, credit card numbers, or other confidential information by observing a victim's screen or keyboard.

related terms: Social Engineering, Shoulder Surfing, Privacy

Packet Analyzer

definition: A software tool that captures and analyzes network traffic.

explanation: It's like a microscope for network traffic, allowing you to see the individual data packets that make up a communication.

real-world examples: Wireshark and tcpdump are popular packet analyzers used by network administrators and security professionals.

related terms: Network Monitoring, Packet Sniffing, Protocol Analysis

Packet Capture

definition: The process of recording network traffic for analysis.

explanation: It's like recording a conversation so you can listen to it later.

real-world examples: Packet capture tools can be used to troubleshoot network problems, detect security threats, or gather evidence of a crime.

related terms: Packet Sniffer, Network Monitoring, Traffic Analysis

Packet Filtering

definition: A technique used by firewalls to control network traffic based on source and destination IP addresses, ports, and protocols.

explanation: It's like a security guard checking IDs and only allowing certain people into a building.

real-world examples: Packet filtering firewalls can block traffic from specific IP addresses, ports, or protocols.

related terms: Firewall, Network Security, Access Control

Packet Injection

definition: The act of inserting forged or modified packets into a network.

explanation: It's like sending a fake message into a conversation to mislead or confuse the participants.

real-world examples: Can be used in various attacks, such as ARP spoofing, DNS poisoning, and denial-of-service (DoS) attacks.

related terms: Packet Forgery, Spoofing, Man-in-the-Middle Attack

Packet Loss

definition: The phenomenon of data packets failing to reach their destination on a network.

explanation: It's like a letter getting lost in the mail - the sender doesn't know if the recipient received it.

real-world examples: Can be caused by network congestion, faulty hardware, or malicious activity.

related terms: Network Performance, Quality of Service (QoS), Network Troubleshooting

Packet Sniffer

definition: A software or hardware tool that intercepts and analyzes network traffic.

explanation: It's like a wiretap for your network, allowing you to see all the data that is being transmitted.

real-world examples: Used for troubleshooting network problems, monitoring network performance, and detecting security threats.

related terms: Packet Capture, Network Monitoring, Traffic Analysis

Parameter Pollution

definition: An attack that exploits a web application's handling of parameters to bypass security controls or manipulate the application's behavior.

explanation: It's like sending a confusing message to a computer, causing it to become disoriented and make mistakes.

real-world examples: Attackers can use parameter pollution to bypass input validation, inject malicious code, or access hidden resources.

related terms: Web Application Security, Injection Attack, Input Validation

Passive Reconnaissance

definition: The process of gathering information about a target without actively engaging with it.

explanation: It's like observing a target from a distance, looking for clues about its vulnerabilities without raising any alarms.

real-world examples: Searching public records, social media profiles, and company websites to gather information.

related terms: Reconnaissance, Open-Source Intelligence (OSINT), Threat Intelligence

Password Aging

definition: A security policy that requires users to change their passwords after a certain period.

explanation: Imagine a carton of milk that expires after a certain date. Password aging works similarly to ensure passwords don't become stale and vulnerable to cracking.

real-world examples: Companies might require employees to change their passwords every 90 days.

related terms: Password Policy, Password Security, Account Lockout

Password Authentication Protocol (PAP)

definition: A simple authentication protocol where usernames and passwords are sent in cleartext over a network.

explanation: Like sending a postcard with your login details written on it - anyone who intercepts it can read it.

real-world examples: PAP is generally considered insecure and is rarely used in modern networks.

related terms: Authentication Protocol, CHAP, Password Security

Password Brute Forcing

definition: A method of cracking passwords by systematically trying every possible combination of characters.

explanation: Imagine trying to unlock a combination lock by trying every possible combination until you find the right one.

real-world examples: Attackers use automated tools to try millions of combinations per second.

related terms: Password Cracking, Dictionary Attack, Brute-Force Attack

Password Complexity

definition: A measure of how difficult a password is to guess.

explanation: It's like the difference between a simple four-digit PIN and a complex password with letters, numbers, and symbols.

real-world examples: Strong passwords should be long, complex, and difficult to guess.

related terms: Password Strength, Password Policy, Authentication

Password Cracking

definition: The process of attempting to recover passwords from data that has been stored in or transmitted by a computer system.

explanation: It's like a thief trying to break into a safe by figuring out the combination.

real-world examples: Hackers use various techniques to crack passwords, such as brute-force attacks, dictionary attacks, and social engineering.

related terms: Password Security, Brute Force Attack, Hash

Password Expiration

definition: A security policy that requires users to change their passwords periodically.

explanation: It's like a driver's license that needs to be renewed every few years to ensure it's still valid.

real-world examples: Companies often have password expiration policies that require employees to change their passwords every 90 or 180 days.

related terms: Password Aging, Password Policy, Account Management

Password Guessing

definition: A simple method of password cracking that involves trying to guess a password based on information about the user.

explanation: It's like trying to guess someone's PIN based on their birthdate or pet's name.

real-world examples: Often used in conjunction with other techniques, such as social engineering or phishing.

related terms: Password Cracking, Brute Force Attack, Social Engineering

Password Hash

definition: A fixed-length string of characters generated from a password using a hash function.

explanation: It's like a unique fingerprint for a password. The hash function transforms the password into a scrambled string that cannot be reversed to reveal the original password.

real-world examples: Password hashes are stored in databases instead of plain text passwords, making them more secure.

related terms: Hash Function, Password Storage, Salting

Password History

definition: A record of a user's previous passwords.

explanation: It's like a logbook of all the passwords you've used in the past.

real-world examples: Used to prevent users from reusing old passwords.

related terms: Password Policy, Password Security, Account Management

Password Length

definition: The number of characters in a password.

explanation: It's like the height of a fence - the taller the fence, the harder it is to climb over.

real-world examples: Longer passwords are generally considered more secure than shorter ones.

related terms: Password Strength, Password Complexity, Password Policy

Password Manager

definition: A software application that stores and manages user passwords.

explanation: It's like a digital keychain that securely stores all your passwords, so you don't have to remember them all.

real-world examples: LastPass, 1Password, and Dashlane are popular password managers.

related terms: Password Security, Encryption, Two-Factor Authentication (2FA)

Password Policy

definition: A set of rules that govern how passwords are created and used.

explanation: It's like a set of house rules for passwords, specifying requirements for length, complexity, and expiration.

real-world examples: Password policies often require users to use a combination of letters, numbers, and symbols, and to change their passwords regularly.

related terms: Password Security, Password Complexity, Password Expiration

Password Reset

definition: The process of changing or recovering a forgotten password.

explanation: It's like getting a new key made for a lock you've lost the key to.

real-world examples: Typically involves answering security questions, receiving a reset code via email or SMS, or contacting customer support.

related terms: Account Recovery, Password Management, Authentication

Password Spraying

definition: An attack that attempts to access many accounts using a small number of commonly used passwords.

explanation: It's like a burglar trying the same key on multiple doors, hoping to find one that's unlocked.

real-world examples: Attackers often target accounts with weak or default passwords.

related terms: Brute Force Attack, Credential Stuffing, Password Security

Password Vault

definition: A secure storage location for passwords, often encrypted and protected with a master password.

explanation: It's like a safe deposit box for your passwords, keeping them secure and accessible only to authorized users.

real-world examples: Password managers often use password vaults to store user credentials.

related terms: Password Manager, Password Security, Encryption

Patch

definition: A piece of software designed to update a computer program or its supporting data, fixing or improving it.

explanation: It's like a band-aid for software - it fixes bugs, vulnerabilities, and other problems.

real-world examples: Software vendors regularly release patches to address security vulnerabilities and improve the functionality of their products.

related terms: Security Update, Software Update, Vulnerability Management

Patch Management

definition: The process of planning, testing, deploying, and verifying software patches.

explanation: It's like a maintenance schedule for your software, ensuring that it's always up-to-date and secure.

real-world examples: Implementing a patch management process to ensure that all systems are patched in a timely manner.

related terms: Vulnerability Management, Software Update, Security

Patch Tuesday

definition: An informal term used to refer to the day of the week (typically the second Tuesday of the month) when Microsoft releases its security patches.

explanation: It's like a monthly reminder to update your windows computer.

real-world examples: Organizations often schedule their patch management activities around Patch Tuesday to ensure that their systems are protected against the latest vulnerabilities.

related terms: Patch, Security Update, Microsoft

Payload

definition: The part of a malware or exploit that performs the intended malicious action.

explanation: It's like the explosive material inside a bomb - it's the part that causes the damage.

real-world examples: The payload of a ransomware attack might be the code that encrypts your files, or the payload of a phishing email might be a link to a malicious website.

related terms: Malware, Exploit, Attack

Payment Card Industry Data Security Standard (PCI DSS)

definition: A set of security standards designed to assure a secure environment is maintained in all companies that accept, process, store, or transmit credit card information.

explanation: It's like a rulebook for businesses that handle credit cards, outlining the steps they need to take to protect their customers' financial information.

real-world examples: Requirements include installing firewalls, encrypting cardholder data, and regularly monitoring and testing networks.

related terms: Credit Card Security, Compliance, Data Security Standard

Penetration Testing (Pen Test)

definition: An authorized simulated cyberattack on a computer system to evaluate the security of the system.

explanation: It's like hiring an ethical hacker to try and break into your house to find any security weaknesses.

real-world examples: This helps organizations find and fix vulnerabilities before malicious hackers can exploit them.

related terms: Ethical Hacking, Vulnerability Assessment, Security Assessment

Perimeter Defense

definition: A security strategy focused on protecting the boundaries of a network.

explanation: Think of it like a castle's walls and moat – the first line of defense against intruders.

real-world examples: Firewalls, intrusion detection/prevention systems (IDS/IPS), and web application firewalls (WAFs) are commonly used perimeter defense tools.

related terms: Network Security, Firewall, Intrusion Detection/Prevention

Perimeter Network

definition: A network segment sitting between an organization's internal network and the internet to act as a buffer zone.

explanation: It's like a moat around a castle, providing an extra layer of protection.

real-world examples: Also known as a demilitarized zone (DMZ), it typically hosts publicly accessible servers like web and email servers.

related terms: DMZ, Network Security, Bastion Host

Perimeter Security

definition: The collective security measures deployed at the perimeter of a network to protect it from external threats.

explanation: It's like the security guards, fences, and cameras that protect a physical building.

real-world examples: Includes firewalls, intrusion detection/prevention systems (IDS/IPS), and other network security technologies.

related terms: Network Security, Firewall, Intrusion Detection/Prevention

Personally Identifiable Information (PII)

definition: Any data that could potentially identify a specific individual.

explanation: It's like a digital fingerprint - it uniquely identifies you and can be used to track your online activity or steal your identity.

real-world examples: Includes your name, address, phone number, email address, Social Security number, and other sensitive information.

related terms: Privacy, Data Protection, Identity Theft

Personally Identifiable Information (PII) Breach

definition: An incident that results in the unauthorized access, disclosure, or theft of personally identifiable information (PII).

explanation: It's like someone stealing your wallet and gaining access to your personal information.

real-world examples: A hacker accessing a company's customer database and stealing credit card numbers, or a government agency losing track of sensitive personal records.

related terms: Data Breach, Identity Theft, Cybercrime

Pharming

definition: A cyber attack that redirects a website's traffic to a fake website that looks identical to the legitimate site.

explanation: Imagine typing in your bank's website address but being taken to a fake website designed to steal your login credentials.

real-world examples: Attackers often compromise DNS servers to redirect traffic to their fake websites.

related terms: Phishing, DNS Poisoning, Spoofing

Phishing

definition: A type of cyber attack that uses fraudulent emails, messages, or websites to trick victims into revealing sensitive information.

explanation: It's like a fisherman casting a line with bait, hoping to hook unsuspecting victims.

real-world examples: Emails that appear to be from your bank asking you to update your password, or fake websites that mimic legitimate ones to steal your login credentials.

related terms: Social Engineering, Spear Phishing, Vishing

Phreaking

definition: The act of hacking into telecommunication systems, especially to make free long-distance calls.

explanation: It's like rewiring a phone system to make free calls or eavesdrop on conversations.

real-world examples: Mostly associated with the early days of hacking, phreaking is now less common due to advances in telecommunications security.

related terms: Hacking, Phone Hacking, Telecommunications Fraud

Physical Access Control

definition: Security measures that restrict physical access to a location or facility.

explanation: It's like the locks, security cameras, and guards that protect a building from unauthorized entry.

real-world examples: Keycard access systems, biometric scanners, fences, and security guards.

related terms: Physical Security, Access Control, Perimeter Security

Physical Security

definition: The protection of personnel, hardware, software, networks, and data from physical actions and events that could cause serious loss or damage to an enterprise, agency, or institution.

explanation: It's like the walls, doors, locks, and security guards that protect a building from physical threats.

real-world examples: Includes measures like locks, alarms, security cameras, and security guards.

related terms: Cybersecurity, Information Security, Access Control

Physical Security Breaches

definition: Unauthorized access to physical locations, equipment, or documents.

explanation: It's like breaking into a building to steal valuable items.

real-world examples: Thieves stealing hardware containing sensitive data from an office.

related terms: Physical Security, Cybersecurity, Access Control

Physical Security Information Management (PSIM)

definition: Software that integrates information from various physical security systems, such as video surveillance, access control, and intrusion detection systems.

explanation: It's like a central command center for physical security, allowing operators to monitor and manage multiple systems from a single interface.

real-world examples: Used to improve situational awareness, streamline security operations, and automate incident response.

related terms: Physical Security, Security Information and Event Management (SIEM), Convergence

Piggybacking

definition: The act of following someone who has legitimate access to gain unauthorized access to a restricted area.

explanation: It's like sneaking into a concert by following someone with a ticket through the entrance.

real-world examples: An attacker might follow an employee through a secure door without showing their own credentials.

related terms: Tailgating, Social Engineering, Physical Security

PII Scanning

definition: The process of searching through data to identify personally identifiable information (PII).

explanation: It's like a detective looking for clues in a document that could identify a person.

real-world examples: Scanning emails, documents, and databases for PII such as names, addresses, social security numbers, or credit card numbers.

related terms: Data Discovery, Data Loss Prevention (DLP), Privacy

Ping Flood

definition: A type of Denial of Service (DoS) attack that overwhelms a target with ICMP echo requests.

explanation: Imagine flooding a store with so many customers that no one else can enter. A ping flood attack works similarly, overwhelming a system with requests to make it unavailable to legitimate traffic.

real-world examples: Used to disrupt online services, websites, and networks.

related terms: DoS Attack, DDoS Attack, ICMP

Ping of Death

definition: A type of DoS attack that sends a malformed or oversized ping packet to crash the target system.

explanation: It's like sending a package that's too big for a mailbox, causing it to break.

real-world examples: This attack exploits vulnerabilities in how systems handle oversized packets.

related terms: DoS Attack, Buffer Overflow, Network Attack

Ping Sweep

definition: A network scanning technique used to discover active hosts on a network by sending ICMP echo requests (pings).

explanation: It's like knocking on doors in a neighborhood to see who's home.

real-world examples: Used by network administrators to map out a network, and by attackers to identify potential targets.

related terms: Network Scanning, Nmap, Reconnaissance

Pivoting

definition: The technique of using an already compromised system to attack other systems on the same network.

explanation: It's like a burglar using one room in a house as a base to break into other rooms.

real-world examples: An attacker might compromise a low-security server and then use it to attack other, more valuable systems.

related terms: Lateral Movement, Privilege Escalation, Persistence

Plaintext

definition: Data that is not encrypted and can be read by anyone.

explanation: It's like sending a postcard - anyone who sees it can read the message.

real-world examples: Plaintext passwords, emails, or documents are easily readable without any special tools or keys.

related terms: Ciphertext, Encryption, Cryptography

Plaintext Attack

definition: A cryptanalysis attack where the attacker has access to the plaintext (unencrypted) version of a message.

explanation: It's like trying to figure out a secret code when you already know what the original message says.

real-world examples: Attackers can use known plaintext attacks to try and deduce the encryption key or algorithm.

related terms: Cryptanalysis, Encryption, Ciphertext

Platform as a Service (PaaS)

definition: A cloud computing model providing a platform for developing, running, and managing applications without the complexity of building and maintaining the infrastructure.

explanation: It's like renting a kitchen to cook a meal - you don't have to worry about building the kitchen or buying the appliances, you just focus on cooking.

real-world examples: Heroku, Google App Engine, and AWS Elastic Beanstalk.

related terms: Cloud Computing, Software as a Service (SaaS), Infrastructure as a Service (IaaS)

Plug-and-Play Attack

definition: An attack that exploits the automatic configuration features of Plug and Play (PnP) devices.

explanation: It's like a stranger plugging a device into your computer without your permission and taking control of it.

real-world examples: USB devices with malicious firmware can be used to install malware or steal data when plugged into a computer.

related terms: USB Attack, BadUSB, Malware

Point-of-sale (POS) Malware

definition: Malware that targets point-of-sale systems to steal payment card information.

explanation: It's like installing a skimmer on a cash register.

real-world examples: Hackers infecting POS terminals in retail stores to capture credit card data.

related terms: Cybercrime, Data Theft, Malware

Point-to-Point Protocol (PPP)

definition: A data link protocol commonly used to establish a direct connection between two networking nodes.

explanation: It's like a direct phone line between two people – it establishes a secure, private connection for data transmission.

real-world examples: Used in dial-up internet connections and some VPNs.

related terms: Network Protocol, Data Link Layer, VPN

Poisoned NULL Byte

definition: A type of code injection attack that inserts a NULL character (%00) into a string to prematurely terminate it.

explanation: It's like putting a period in the middle of a sentence, causing the rest of the sentence to be ignored.

real-world examples: Attackers can use poisoned NULL bytes to bypass input validation and execute malicious codes.

related terms: Injection Attack, Input Validation, Web Application Security

Policy-Based Access Control (PBAC)

definition: An access control method that uses policies to determine access permissions.

explanation: It's like setting rules for who can access what based on policies rather than individual permissions.

real-world examples: Granting access to company resources based on job roles or project requirements and adjusting access automatically as roles change.

related terms: Role-Based Access Control (RBAC), Attribute-Based Access Control (ABAC), Access Management, Security Policies

Polymorphic Malware

definition: Malware that can change its code to evade detection by antivirus software.

explanation: It's like a chameleon that changes its colors to blend in with its surroundings.

real-world examples: Polymorphic malware uses techniques like encryption and code obfuscation to make each copy of itself look different.

related terms: Malware, Antivirus Evasion, Metamorphic Malware

Port

definition: A communication endpoint for a specific service or application on a computer or network device.

explanation: It's like a numbered mailbox for your computer - each port receives a different type of message.

real-world examples: Web servers typically use port 80, email servers use port 25, and SSH servers use port 22.

related terms: Network Protocol, IP Address, Firewall

Port Forwarding

definition: A technique allowing external devices to access specific ports on a private network.

explanation: It's like forwarding your mail to a different address.

real-world examples: Used to access devices like security cameras or web servers on a home network from the internet.

related terms: Network Address Translation (NAT), Router, Remote Access

Port Honeypot

definition: A decoy port that is intentionally left open to attract and monitor attackers.

explanation: It's like leaving a fake safe in a bank vault to see if anyone tries to break into it.

real-world examples: Used to gather information about attacker tactics and techniques, and to detect and respond to attacks.

related terms: Honeypot, Deception Technology, Intrusion Detection

Port Knocking

definition: A method of externally opening ports on a firewall by generating a connection attempt on a set of prespecified closed ports.

explanation: It's like a secret knock that you use to signal to a friend that you want to be let in.

real-world examples: Only when the correct sequence of knocks is received, the firewall will open the desired port.

related terms: Firewall, Port, Security

Port Mirroring

definition: A networking technique that sends a copy of network packets seen on one port (or an entire VLAN) to a network monitoring connection on another port.

explanation: It's like having a security camera that records all the traffic passing through a particular intersection.

real-world examples: Used to monitor network traffic for security and troubleshooting purposes.

related terms: Network Monitoring, Packet Sniffing, Traffic Analysis

Port Scanning

definition: The process of sending packets to different ports on a computer or network to determine which ones are open.

explanation: It's like knocking on doors in a neighborhood to see who's home.

real-world examples: Hackers use port scanning to identify potential targets, while security professionals use it to assess the security of their networks.

related terms: Network Scanning, Nmap, Vulnerability Assessment

Port Security

definition: A security feature that limits the number of MAC addresses allowed to connect to a switch port.

explanation: It's like having a guest list for your Wi-Fi network – only devices on the list are allowed to connect.

real-world examples: Used to prevent unauthorized devices from accessing a network.

related terms: MAC Address Filtering, Network Security, Switch

Port Triggering

definition: A method of automatically opening a specific port on a firewall when a certain type of traffic is detected.

explanation: It's like a motion-activated light that turns on when someone approaches your house.

real-world examples: Used to allow incoming traffic for specific applications, such as online games or video conferencing.

related terms: Port Forwarding, Firewall, Network Security

Pretexting

definition: A form of social engineering in which an individual lies to obtain privileged data.

explanation: It's like a con artist creating a fake story to gain someone's trust.

real-world examples: Attackers might pose as IT support to trick employees into giving them their passwords, or as a potential customer to obtain sensitive company information.

related terms: Social Engineering, Phishing, Impersonation

Pretty Good Privacy (PGP)

definition: A data encryption and decryption computer program that provides cryptographic privacy and authentication for data communication.

explanation: It's like a super-secret decoder ring for your emails. It encrypts your messages so only the intended recipient can read them.

real-world examples: Used for secure email communication, file encryption, and digital signatures.

related terms: Encryption, Decryption, GPG (GNU Privacy Guard)

Printer Hacking

definition: Exploiting vulnerabilities in printers to gain unauthorized access to networks or steal data.

explanation: It's like tapping into a phone line to eavesdrop on conversations.

real-world examples: Hackers gaining access to a network through an insecure printer.

related terms: Network Security, Cybersecurity, IoT Security

Privacy

definition: The state of being free from observation or disturbance of one's personal life.

explanation: It's like having a personal bubble that protects you from unwanted attention.

real-world examples: Online privacy concerns include tracking, data collection, and targeted advertising.

related terms: Data Protection, Confidentiality, Personally Identifiable Information (PII)

Privacy by Design

definition: An approach to systems engineering that takes privacy into account throughout the whole engineering process.

explanation: It's like building a house with security features included from the very beginning, rather than adding them later.

real-world examples: Designing a new app that collects the minimum amount of personal data necessary, and ensuring it is encrypted.

related terms: Data Protection, Privacy Engineering, GDPR, Security by Design

Privacy Impact Assessment (PIA)

definition: An analysis of how a project or system might affect the privacy of individuals.

explanation: It's like a checklist for privacy, ensuring that new technologies or processes don't violate people's personal information.

real-world examples: Conducted before implementing new systems or collecting personal data.

related terms: Data Protection, GDPR, Privacy by Design

Private Branch Exchange (PBX) Hacking

definition: The act of gaining unauthorized access to a company's phone system.

explanation: It's like a burglar tapping into your phone lines to listen in on conversations or make free calls.

real-world examples: Hackers might use PBX hacking to steal sensitive information, eavesdrop on calls, or reroute calls to premium-rate numbers.

related terms: Telecommunications Fraud, Hacking, Voicemail Hacking

Private Key

definition: A cryptographic key that is kept secret and used to decrypt messages or verify digital signatures.

explanation: It's like the key to a safe deposit box – only the person with the private key can access the contents.

real-world examples: Used in asymmetric encryption to decrypt messages that were encrypted with the corresponding public key.

related terms: Public Key, Asymmetric Encryption, Cryptography

Privilege

definition: A special right, advantage, or immunity granted or available only to a particular person or group of people.

explanation: In cybersecurity, privileges are the permissions granted to users, processes, or programs to access or modify resources.

real-world examples: Administrator privileges, root access, and elevated permissions.

related terms: Access Control, Authorization, Least Privilege

Privilege Creep

definition: The gradual accumulation of access rights beyond what is required for an individual's job function.

explanation: It's like someone slowly collecting keys to different rooms in a building, even though they only need access to a few.

real-world examples: A user who starts as an intern might gradually gain more privileges over time, even if their role doesn't change.

related terms: Access Management, Least Privilege, Segregation of Duties

Privilege Escalation

definition: The act of exploiting a bug, design flaw, or configuration oversight in an operating system or software application to gain elevated access to resources that are normally protected from an application or user.

explanation: It's like a thief finding a way to pick the lock on a door that was supposed to be secure.

real-world examples: Attackers might exploit a vulnerability in a web application to gain access to the underlying operating system.

related terms: Vulnerability, Exploit, Rootkit

Processor Security

definition: Hardware and software technologies designed to protect the processor from attacks and unauthorized access.

explanation: It's like a security system for the brain of your computer, ensuring that it can't be tampered with or manipulated.

real-world examples: Hardware-based security features like Intel's SGX (Software Guard Extensions) and AMD's SEV (Secure Encrypted Virtualization) are designed to protect the processor from attacks.

related terms: Hardware Security, Trusted Computing, Secure Boot

Protected Health Information (PHI)

definition: Any information about health status, provision of health care, or payment for health care that can be linked to a specific individual.

explanation: It's like your medical records - it contains sensitive information about your health that needs to be protected.

real-world examples: Names, addresses, social security numbers, medical diagnoses, and treatment information.

related terms: HIPAA, Healthcare IT, Privacy

Protocol

definition: A set of rules or procedures for transmitting data between electronic devices.

explanation: It's like the rules of grammar that govern how we construct sentences - protocols define how computers communicate with each other.

real-world examples: TCP/IP, HTTP, FTP, SMTP.

related terms: Networking, Internet, Communication

Protocol Analysis

definition: The process of capturing and analyzing network traffic to understand how protocols are used and to identify potential security threats.

explanation: It's like listening in on a conversation between two computers to see what they're talking about.

real-world examples: Used to troubleshoot network problems, detect intrusions, and analyze malware traffic.

related terms: Packet Sniffing, Network Monitoring, Wireshark

Protocol Analyzer

definition: A software tool capturing and analyzing network traffic.

explanation: It's like a microscope for network traffic, allowing you to see the individual data packets that make up a communication.

real-world examples: Wireshark and tcpdump are popular protocol analyzers.

related terms: Packet Sniffing, Network Monitoring, Traffic Analysis

Proxy

definition: A server application or appliance acting as an intermediary for requests from clients seeking resources from servers that provide those resources.

explanation: It's like a middleman between you and the website you're trying to access.

real-world examples: Web proxies can be used to filter content, cache websites, or bypass restrictions.

related terms: Proxy Server, Web Proxy, Anonymizer

Proxy Server

definition: A server acting as an intermediary for requests from clients seeking resources from other servers.

explanation: It's like a receptionist who forwards calls to the appropriate person in an office.

real-world examples: Used to filter content, cache websites, and provide anonymity.

related terms: Proxy, Web Proxy, Anonymizer

Pseudonymization

definition: The processing of personal data in such a manner that the personal data can no longer be attributed to a specific data subject without the use of additional information.

explanation: It's like using a pen name instead of your real name.

real-world examples: Replacing names with unique identifiers or masking email addresses.

related terms: Data Protection, Privacy, Anonymization

Pseudorandom Number Generator (PRNG)

definition: An algorithm that generates a sequence of numbers that appears random but is actually deterministic.

explanation: It's like a magic trick that produces seemingly random numbers, but the magician knows the secret behind the trick.

real-world examples: Used in cryptography for generating keys, nonces, and other random values.

related terms: Random Number Generator (RNG), Cryptography, Security

Public Key

definition: A cryptographic key that can be openly distributed and is used to encrypt messages or verify digital signatures.

explanation: It's like a public mailbox that anyone can use to send you a message, but only you have the key to open it.

real-world examples: Used in asymmetric encryption to encrypt messages that can only be decrypted with the corresponding private key.

related terms: Private Key, Asymmetric Encryption, Cryptography

Public Key Cryptography

definition: A cryptographic system that uses pairs of keys: public keys, which may be disseminated widely, and private keys, which are known only to the owner.

explanation: It's like having a public mailbox for receiving messages and a private key for opening it.

real-world examples: Used for secure email communication, digital signatures, and secure web browsing (HTTPS).

related terms: Asymmetric Encryption, Private Key, Cryptography

Public Key Cryptography Standards (PKCS)

definition: A set of standards that define the technical details of how to use public key cryptography.

explanation: It's like a set of instructions for using a public key cryptography system, ensuring that different systems can communicate securely.

real-world examples: PKCS standards cover topics like key generation, encryption, and digital signatures.

related terms: Public Key Cryptography, Cryptography, Security Standards

Public Key Infrastructure (PKI)

definition: A system for creating, storing, and distributing digital certificates that are used to verify the ownership of public keys.

explanation: Think of it like a passport system for the digital world. PKI uses digital certificates to establish trust between parties, ensuring secure communication and transactions.

real-world examples: Used in SSL/TLS for secure web browsing, email encryption, and code signing.

related terms: Digital Certificate, Certificate Authority (CA), Encryption

Puffer

definition: A piece of code intentionally inserted into a software program to make it harder to reverse engineer.

explanation: It's like adding extra fluff to a jacket to make it harder to take apart and figure out how it was made.

real-world examples: Software developers may use puffers to protect their intellectual property from being copied.

related terms: Code Obfuscation, Anti-Reverse Engineering

Purple Team

definition: A security team that combines the skills and knowledge of red teams (attackers) and blue teams (defenders).

explanation: It's like a sports team with both offensive and defensive players - they work together to find and fix vulnerabilities before attackers can exploit them.

real-world examples: Purple teams collaborate to improve an organization's overall security posture by sharing knowledge and expertise.

related terms: Red Team, Blue Team, Threat Hunting

Push Button Reset Attack

definition: An attack that exploits a vulnerability in IoT devices with physical reset buttons to reset the device to factory defaults, potentially bypassing security settings.

explanation: Imagine a burglar pressing the reset button on your smart home hub to disable the alarm system.

real-world examples: Attackers can use this technique to gain access to devices and networks.

related terms: IoT Security, Physical Security, Default Credentials

Pwned

definition: Slang term for "owned," meaning to be defeated or controlled, often used in the context of a system being compromised.

explanation: It's like having your house taken over by an intruder.

real-world examples: A user discovering their email account has been compromised in a data breach.

related terms: Hacked, Compromised, Cybersecurity

QRishing

definition: A type of phishing attack that uses QR codes to deliver malicious payloads.

explanation: It's like a QR code that, when scanned, takes you to a fake website designed to steal your information.

real-world examples: Attackers might place malicious QR codes on posters, flyers, or even product packaging.

related terms: Phishing, Social Engineering, QR Code

Quantum Computing

definition: A type of computing that uses the principles of quantum mechanics to perform calculations.

explanation: Traditional computers use bits, which can be either 0 or 1. Quantum computers use qubits, which can be 0, 1, or both at the same time.

real-world examples: Quantum computers could potentially break many of the encryption algorithms used today, posing a significant threat to cybersecurity.

related terms: Quantum Cryptography, Post-Quantum Cryptography, Encryption

Quantum Cryptography

definition: A type of cryptography that uses the principles of quantum mechanics to secure communications.

explanation: It's like using the laws of physics to create unbreakable codes.

real-world examples: Quantum key distribution (QKD) is a quantum cryptography technique that allows two parties to securely share encryption keys.

related terms: Quantum Computing, Encryption, Cryptography

Quantum Cryptography Attacks

definition: Attacks that exploit the principles of quantum computing to break traditional encryption methods.

explanation: It's like using a futuristic device to crack today's security codes.

real-world examples: Potential future attacks using quantum computers to decrypt secure communications.

related terms: Quantum Computing, Cryptography, Cybersecurity

Quarantine

definition: The isolation of a file, computer, or network that is suspected of being infected with malware.

explanation: It's like putting a sick person in quarantine to prevent them from infecting others.

real-world examples: Antivirus software might quarantine infected files to prevent them from causing harm.

related terms: Malware, Antivirus, Containment

Quarantine Network

definition: A separate network segment used to isolate infected or suspicious devices.

explanation: It's like a hospital ward for infected computers, where they can be treated without risking the health of other devices.

real-world examples: Used to prevent malware from spreading to other parts of the network.

related terms: Network Segmentation, Isolation, Malware Containment

Query Parameterization

definition: A technique used to prevent SQL injection attacks by treating user input as data, rather than executable code.

explanation: It's like putting quotation marks around a word in a sentence to indicate that it's not a command.

real-world examples: Web applications use query parameterization to safely incorporate user input into SQL queries, preventing attackers from injecting malicious code.

related terms: SQL Injection, Input Validation, Web Application Security

Query String

definition: The part of a URL that contains parameters or variables.

explanation: It's like the instructions you give a website when you click on a link.

real-world examples: A query string might specify the search terms you entered or the page you want to view.

related terms: URL, Web Application, Parameter

Quid Pro Quo

definition: A type of social engineering attack where an attacker offers a benefit in exchange for information or access.

explanation: It's like a bribe – the attacker promises something valuable in return for your cooperation.

real-world examples: An attacker might offer free software or a gift card in exchange for your login credentials.

related terms: Social Engineering, Bribery, Deception

Race Condition

definition: A software vulnerability that occurs when the behavior of a program depends on the timing of events.

explanation: It's like two people trying to grab the same seat on a bus – whoever gets there first wins.

real-world examples: Race conditions can be exploited by attackers to gain unauthorized access or cause a system to crash.

related terms: Software Bug, Concurrency, Timing Attack

RADIUS (Remote Authentication Dial-In User Service)

definition: A networking protocol that provides centralized authentication, authorization, and accounting (AAA) management for users who connect and use a network service.

explanation: It's like a hotel concierge who checks your ID, assigns you a room, and keeps track of your phone calls and room service charges.

real-world examples: Used by internet service providers (ISPs), mobile networks, and enterprise networks to manage user access.

related terms: Authentication, Authorization, Accounting (AAA)

Rainbow Table Attack

definition: A type of password attack that uses a rainbow table to crack password hashes.

explanation: It's like using a dictionary to look up the meaning of a word – the attacker compares the hash of a stolen password to the hashes in the rainbow table to find the original password.

real-world examples: This type of attack is effective against weak passwords but can be mitigated by using strong password hashing algorithms and salting.

related terms: Password Cracking, Rainbow Table, Hash Function

RAM Scraper

definition: Malware that searches the computer's RAM for sensitive data, such as credit card numbers or passwords.

explanation: It's like a thief rummaging through your wallet while you're not looking.

real-world examples: RAM scrapers are often used in point-of-sale (POS) system attacks to steal credit card data.

related terms: Malware, Data Theft, Memory Forensics

Random Number Generator (RNG)

definition: A device or algorithm that creates a sequence of numbers that lacks any pattern.

explanation: It's like rolling a die – each number that comes up is unpredictable.

real-world examples: Used in cryptography for generating encryption keys, nonces, and other random values.

related terms: Pseudorandom Number Generator (PRNG), Cryptography, Security

Ransomware

definition: Malware that encrypts a victim's files and demands a ransom payment to decrypt them.

explanation: It's like a digital kidnapping – your files are held hostage until you pay up.

real-world examples: CryptoLocker, WannaCry, and Ryuk are examples of ransomware.

related terms: Malware, Encryption, Crypto-Malware

Ransomware as a Service (RaaS)

definition: A business model where cybercriminals offer ransomware and related services to other criminals for a fee.

explanation: It's like a franchise for cybercrime, allowing less-skilled attackers to launch ransomware attacks without having to develop their own malware or infrastructure.

real-world examples: DarkSide, REvil, and GandCrab are examples of RaaS operations.

related terms: Ransomware, Malware, Cybercrime

Ransomware Negotiation

definition: The process of communicating with ransomware attackers to potentially lower the ransom demand or obtain a decryption key without paying.

explanation: It's like negotiating with a kidnapper, but instead of a person, it's your data that's being held hostage.

real-world examples: Often involves engaging with professional negotiators or law enforcement agencies to handle communications with the attackers.

related terms: Ransomware, Cyber Insurance, Incident Response

Real-Time Blackhole List (RBL)

definition: A list of IP addresses known to send spam or engage in other malicious activities.

explanation: It's like a "do not fly" list for airplanes but for IP addresses. Email servers use RBLs to block incoming emails from these addresses.

real-world examples: Spamhaus, Spamcop, and Barracuda Central maintain RBLs that are used by email providers and organizations to block spam.

related terms: Blacklist, Email Filtering, Spam

Reconnaissance

definition: The process of gathering information about a target before launching an attack.

explanation: It's like a spy gathering intelligence on a target before a mission. Attackers use reconnaissance to identify vulnerabilities and plan their attacks.

real-world examples: Scanning networks for open ports, researching a company's online presence, and using social engineering to gather information from employees.

related terms: Active Reconnaissance, Passive Reconnaissance, Threat Intelligence

Red Team

definition: A group of security professionals who simulate real-world attacks for testing, testing an organization's defenses.

explanation: Think of them as the "bad guys" in a training exercise, trying to find and exploit weaknesses before real attackers do.

real-world examples: Red teams use penetration testing, social engineering, and other tactics to simulate attacks.

related terms: Penetration Testing, Adversary Emulation, Blue Team

Redirection

definition: The act of sending a user or data to a different location than originally intended.

explanation: It's like a detour sign that leads you down a different road than you expected.

real-world examples: URL redirection, port redirection, and traffic redirection.

related terms: URL Redirection, Port Forwarding, Load Balancing

Refactoring

definition: The process of restructuring existing computer code without changing its external behavior.

explanation: Imagine rearranging the furniture in your house to make it more functional and efficient.

real-world examples: Refactoring code can improve its readability, maintainability, and performance, and can also help to identify and fix security vulnerabilities.

related terms: Code Review, Software Development, Security

Reflective XSS

definition: A type of cross-site scripting (XSS) attack where the malicious script is reflected from the web server to the user's browser.

explanation: It's like a mirror reflecting a harmful image back at you. The attacker injects malicious code into a website, and the website unknowingly sends the code back to the user's browser, where it is executed.

real-world examples: An attacker might inject a script into a search query that is then reflected back in the search results page, executing in the victim's browser.

related terms: Cross-Site Scripting (XSS), Web Application Security, Injection Attack

Regulatory Compliance

definition: Compliance to laws, regulations, guidelines, and specifications relevant to the business.

explanation: It's like following the rules set by the game officials.

real-world examples: Ensuring data protection practices comply with GDPR.

related terms: Compliance Audit, Risk Management, Governance

Regulatory Reporting

definition: The process of collecting and submitting data that are required by regulatory bodies.

explanation: It's like submitting your homework for review to ensure it meets all the guidelines.

real-world examples: Financial institutions reporting to the SEC.

related terms: Compliance, Risk Management, Audit

Remote Access Trojan (RAT)

definition: Malware that allows an attacker to control a victim's computer remotely.

explanation: Imagine a secret backdoor installed in your computer, giving a stranger complete control over it.

real-world examples: Attackers use RATs to steal data, install additional malware, or spy on the victim's activities.

related terms: Malware, Backdoor, Trojan

Remote Code Execution (RCE)

definition: A vulnerability that lets an attacker execute arbitrary code on a remote system.

explanation: It's like giving a stranger the keys to your car and letting them drive it wherever they want.

real-world examples: Attackers can exploit RCE vulnerabilities to take control of systems, steal data, or launch further attacks.

related terms: Vulnerability, Exploit, Injection Attack

Remote Desktop Protocol (RDP)

definition: A proprietary protocol developed by Microsoft, which provides a user with a graphical interface to connect to another computer over a network connection.

explanation: It's like a remote control for a computer, allowing you to access it from another location.

real-world examples: Commonly used for remote administration and troubleshooting.

related terms: Remote Access, Remote Administration, Virtual Network Computing (VNC)

Remote Desktop Protocol (RDP) Attack

definition: A type of attack that exploits vulnerabilities in the Remote Desktop Protocol (RDP) in order to gain unauthorized access to a system.

explanation: It's like a burglar using a secret passage to enter a house without setting off the alarm.

real-world examples: Attackers can use RDP attacks to install malware, steal data, or launch further attacks.

related terms: Remote Desktop Protocol (RDP), Vulnerability, Brute-Force Attack

Remote File Inclusion (RFI)

definition: A type of web vulnerability that allows an attacker to include a remote file on a server.

explanation: It's like a burglar finding a blueprint for a house that reveals its hidden weaknesses.

real-world examples: Attackers can exploit RFI vulnerabilities to execute malicious code, steal data, or deface websites.

related terms: File Inclusion Vulnerability, Local File Inclusion (LFI), Web Application Security

Remote Procedure Call (RPC)

definition: A protocol that lets a computer program request a service from a program located on another computer on a network without having to understand details of the network.

explanation: It's like calling a friend in another city and asking them to do something for you.

real-world examples: Used for communication between distributed applications and services.

related terms: Client-Server, Distributed Computing, Network Protocol

Replay Attack

definition: An attack where an attacker intercepts a valid data transmission and retransmits it, by impersonating one of the parties.

explanation: It's like playing back a recorded conversation to try and trick someone into thinking you're someone else.

real-world examples: Attackers might replay a captured authentication request to gain unauthorized access to a system.

related terms: Man-in-the-Middle Attack, Packet Sniffing, Authentication

Repudiation

definition: The denial by one of the entities involved in a communication of having participated in all or part of the communication.

explanation: It's like claiming you didn't send a text message, even though there's evidence that you did.

real-world examples: Digital signatures are used to prevent repudiation by providing proof of the sender's identity.

related terms: Non-repudiation, Digital Signature, Authentication

Resilience

definition: The ability of a system or network to withstand and recover from disruptions.

explanation: It's like a rubber band that can be stretched but always returns to its original shape.

real-world examples: Implementing redundancy, failover mechanisms, and disaster recovery plans to ensure that systems can quickly recover from outages or attacks.

related terms: High Availability, Fault Tolerance, Disaster Recovery

Reverse Engineering

definition: The process of analyzing a software or hardware system to understand its design, construction, and operation.

explanation: It's like taking apart a watch to see how it works.

real-world examples: Used to analyze malware, find vulnerabilities in software, and develop compatible products.

related terms: Malware Analysis, Vulnerability Research, Software Development

Reverse Proxy

definition: A type of proxy server that retrieves resources on behalf of a client from one or more servers.

explanation: It's like a personal assistant who fetches things for you.

real-world examples: Used to improve security, performance, and scalability of web applications.

related terms: Proxy Server, Web Server, Load Balancing

Reverse Shell

definition: A type of shell connection established on a connection that is initiated from a remote machine, not from the local host.

explanation: Imagine someone secretly installing a backdoor on your computer that allows them to control it remotely.

real-world examples: Attackers use reverse shells to maintain access to compromised systems and evade detection.

related terms: Shell, Remote Access, Backdoor

RFID Skimming

definition: The unauthorized scanning and copying of data from RFID-enabled cards and devices.

explanation: It's like silently copying the keys to a car without the owner's knowledge.

real-world examples: Thieves using RFID skimmers to steal credit card information from contactless cards.

related terms: Data Theft, Physical Security, Cybersecurity

Risk

definition: The potential for loss, damage, or destruction of an asset as a result of a threat exploiting a vulnerability.

explanation: It's like the chance of getting injured while playing a sport - the more dangerous the sport, the higher the risk.

real-world examples: The risk of a data breach due to a software vulnerability, or the risk of a fire damaging a data center.

related terms: Threat, Vulnerability, Impact

Risk Acceptance

definition: A risk management strategy that involves acknowledging and accepting the potential consequences of a risk without taking any action to mitigate it.

explanation: It's like deciding to drive without car insurance - you're aware of the risk of an accident, but you choose to accept it.

real-world examples: A company might accept the risk of a minor data breach if the cost of implementing security controls is too high.

related terms: Risk Management, Risk Avoidance, Risk Mitigation

Risk Appetite

definition: The level of risk that an organization is willing to accept in pursuit of its objectives.

explanation: It's like a person's tolerance for spicy food - some people can handle a lot of heat, while others prefer milder flavors.

real-world examples: A company with a high-risk appetite might be willing to invest in risky ventures, while a company with a low-risk appetite might be more conservative.

related terms: Risk Management, Risk Tolerance, Risk Assessment

Risk Assessment

definition: The process of identifying, analyzing, and evaluating risks.

explanation: It's like a doctor performing a check-up on a patient to assess their health and identify potential problems.

real-world examples: A risk assessment might involve identifying vulnerabilities in a system, assessing the likelihood of a threat exploiting those vulnerabilities, and estimating the potential impact of a successful attack.

related terms: Risk Management, Threat Assessment, Vulnerability Assessment

Risk Avoidance

definition: A risk management strategy that involves taking steps to eliminate a risk altogether.

explanation: It's like avoiding a dangerous road by taking a different route.

real-world examples: A company might avoid the risk of a fire by storing its data in a fireproof vault.

related terms: Risk Management, Risk Acceptance, Risk Mitigation

Risk Culture

definition: The norms and traditions of behavior within an organization that determine how it identifies, understands, discusses, and acts on risk.

explanation: It's like the personality of the organization regarding risk awareness and management.

real-world examples: Promoting a culture where employees feel comfortable reporting potential risks.

related terms: Risk Management, Governance, Compliance

Risk Governance

definition: The framework through which an organization manages risks to achieve its objectives.

explanation: It's like the command center where strategies for managing risks are developed and monitored.

real-world examples: Establishing a risk management committee to oversee risk policies.

related terms: Governance, Risk Management, Compliance

Risk Heat Map

definition: A visual tool for representing the risks faced by an organization, typically using a matrix to show the likelihood and impact of risks.

explanation: It's like a weather map showing areas of high and low risk.

real-world examples: Creating a heat map to prioritize cybersecurity risks based on their severity and likelihood.

related terms: Risk Assessment, Risk Management, Compliance

Risk Identification

definition: The process of finding, recognizing, and recording risks.

explanation: It's like spotting potential hazards before they become serious issues.

real-world examples: Identifying risks in a new product launch.

related terms: Risk Assessment, Risk Management, Compliance

Risk Management

definition: The process of identifying, assessing, and controlling threats to an organization's capital and earnings.

explanation: It's like having a plan for dealing with unexpected events, such as a fire, a natural disaster, or a cyberattack.

real-world examples: Implementing security controls, purchasing insurance, and developing incident response plans are all examples of risk management activities.

related terms: Risk Assessment, Risk Mitigation, Risk Governance

Risk Mitigation

definition: A risk management strategy that involves taking steps to reduce the probability or impact of a risk.

explanation: It's like wearing a seatbelt when driving - it doesn't eliminate the risk of an accident, but it reduces the severity of the injury if one occurs.

real-world examples: Installing antivirus software, patching vulnerabilities, and implementing access controls are examples of risk mitigation strategies.

related terms: Risk Management, Risk Avoidance, Risk Transference

Risk Register

definition: A document that lists all identified risks, their likelihood and impact, and the mitigation strategies that will be used to address them.

explanation: It's like a to-do list for risk management, keeping track of all the risks that need to be addressed.

real-world examples: The risk register is a living document that is updated as new risks are identified or as the status of existing risks changes.

related terms: Risk Management, Risk Assessment, Risk Mitigation

Risk Tolerance

definition: The amount of risk that an organization is willing to accept.

explanation: It's like a person's pain tolerance - some people can handle more pain than others.

real-world examples: Organizations with a high-risk tolerance might be willing to accept more risk in exchange for potential rewards, while organizations with a low-risk tolerance might be more cautious.

related terms: Risk Appetite, Risk Management, Risk Assessment

Risk Transference

definition: A risk management strategy that involves transferring the risk to another party, such as an insurance company.

explanation: It's like buying insurance for your car - you're transferring the financial risk of an accident to the insurance company.

real-world examples: Purchasing cyber insurance to cover the costs of a data breach is an example of risk transference.

related terms: Risk Management, Risk Avoidance, Cyber Insurance

Risk-Based Authentication

definition: An adaptive authentication method that assesses the risk level of a login attempt and adjusts the required authentication process accordingly.

explanation: It's like asking for more proof of identity if something seems suspicious, like logging in from a new device.

real-world examples: Requiring additional verification for a login attempt from an unfamiliar location or device.

related terms: Multi-Factor Authentication (MFA), Adaptive Authentication, Access Management, Security Policies

Rogue Access Point

definition: A wireless access point that has been installed on a secure network without explicit authorization from a local network administrator whether added by a well-meaning employee or by a malicious attacker.

explanation: Imagine someone setting up a Wi-Fi hotspot in your office without your permission.

real-world examples: Employees might set up rogue access points to improve connectivity, but they can also be used by attackers to intercept traffic or launch attacks.

related terms: Wireless Security, Wi-Fi Pineapple, Evil Twin

Rogue Antivirus

definition: Fake antivirus software that claims to protect your computer but actually installs malware or other unwanted software.

explanation: It's like a wolf in sheep's clothing – it pretends to be helpful, but it's actually harmful.

real-world examples: Rogue antivirus software often uses scare tactics to trick users into installing it.

related terms: Malware, Scareware, Social Engineering

Rogue DHCP Server

definition: A DHCP server on a network that is not under the administrative control of the network staff. It is a network attack that broadcasts forged DHCP replies to all broadcast DHCP discover messages on a network.

explanation: It's like a rogue food vendor setting up shop in a school cafeteria and serving contaminated food.

real-world examples: Rogue DHCP servers can assign incorrect IP addresses or gateway settings, disrupting network traffic or redirecting users to malicious websites.

related terms: DHCP Server, Network Attack, Man-in-the-Middle Attack

Rogue Software

definition: Software that appears legitimate but performs malicious activities.

explanation: It's like a Trojan horse that appears friendly but carries hidden dangers.

real-world examples: Fake antivirus programs that steal user data or install additional malware.

related terms: Malware, Social Engineering, Cybercrime

Root Access

definition: The highest level of access to a computer system, allowing full control over the system.

explanation: It's like having the master key to every lock in a building.

real-world examples: System administrators having root access to manage and configure servers.

related terms: Administrator, Privileged Access, System Control

Root Certificate

definition: A public key certificate that identifies the root certificate authority (CA).

explanation: It's the top-level certificate in a chain of trust, used to verify the authenticity of other certificates.

real-world examples: If you trust a root certificate, you trust all the certificates that it has signed.

related terms: Certificate Authority (CA), Digital Certificate, Public Key Infrastructure (PKI)

Rooting

definition: The process of gaining privileged control (known as root access) over an Android operating system.

explanation: It's like jailbreaking an iPhone - you remove the restrictions imposed by the manufacturer to gain full control over the device.

real-world examples: Allows users to install custom firmware, remove pre-installed apps, and access system files.

related terms: Jailbreaking, Android, Privilege Escalation

Rootkit

definition: A set of software tools allowing an unauthorized user to gain control of a computer system without being detected.

explanation: It's like a secret passage in a castle that allows an intruder to bypass security measures and gain access to the inner workings of the system.

real-world examples: Rootkits are often used by malware to hide their presence and maintain persistence on a system.

related terms: Malware, Stealth, Persistence

Rootkit Scanner

definition: A software tool that searches for and detects rootkits on a computer system.

explanation: It's like a metal detector that can find hidden weapons. In this case, the weapon is a rootkit, and the goal is to find and remove it before it can cause harm.

real-world examples: Rootkit scanners use a variety of techniques to detect rootkits, such as signature scanning, behavioral analysis, and memory analysis.

related terms: Rootkit, Malware Detection, Security Software

Router

definition: A networking device that forwards data packets between computer networks.

explanation: It's like a traffic cop directing cars at an intersection, ensuring that data packets reach their intended destinations.

real-world examples: Home routers connect your home network to the internet, while enterprise routers connect different networks within a company.

related terms: Network, Networking Device, Gateway

Rowhammer Attack

definition: A type of attack that exploits a vulnerability in DRAM (dynamic random-access memory) chips.

explanation: It's like repeatedly slamming a door until the lock breaks. In this case, the attacker repeatedly accesses rows of memory to cause bit flips in adjacent rows, potentially altering data or gaining unauthorized access.

real-world examples: Rowhammer attacks have been used to escalate privileges and bypass security mechanisms.

related terms: Hardware Vulnerability, DRAM, Bit Flip

Rubber Ducky

definition: A USB device that looks like a normal flash drive but is programmed to inject keystrokes into a computer.

explanation: Imagine a tiny robot that plugs into a computer and types commands at lightning speed.

real-world examples: Used by penetration testers and security researchers to simulate keyboard attacks and automate tasks.

related terms: USB Attack, HID Attack, Keystroke Injection

Runtime Application Self-Protection (RASP)

definition: A security technology that is embedded into an application or runtime environment and can detect and prevent attacks in real time.

explanation: It's like having a bodyguard for your application, monitoring its behavior and protecting it from attacks.

real-world examples: RASP can block SQL injection, cross-site scripting (XSS), and other types of attacks.

related terms: Application Security, Web Application Firewall (WAF), Intrusion Prevention System (IPS)

Safe Browsing

definition: A service provided by Google that helps protect users from visiting malicious websites.

explanation: Think of it as a warning sign for dangerous websites, helping you avoid scams, malware, and phishing attacks.

real-world examples: Integrated into Google Chrome, Mozilla Firefox, and Apple Safari.

related terms: Google, Phishing Protection, Malware Protection

Safe Harbor Principles

definition: A legal framework that allows U.S. companies to comply with European Union data protection regulations.

explanation: It's like a bridge between two different sets of privacy laws, allowing companies to transfer data across borders while still protecting it.

real-world examples: Allows U.S. companies to transfer personal data to Europe under certain conditions.

related terms: GDPR, Data Protection, Privacy Shield

Salt

definition: Random data that is added to a password before it is hashed.

explanation: It's like adding a secret ingredient to a recipe to make it more difficult to replicate.

real-world examples: Salting helps to protect against rainbow table attacks by making it harder for attackers to crack password hashes.

related terms: Password Hashing, Rainbow Table Attack, Cryptography

Salted Hash

definition: A hash function that has been applied to a password combined with a salt value.

explanation: It's like adding a layer of encryption to a password hash to make it even more secure.

real-world examples: Salted hashes are more resistant to cracking than unsalted hashes.

related terms: Password Hash, Hash Function, Salting

Sandbox

definition: A secure and isolated environment where programs or files can be executed without affecting the host system.

explanation: It's like a quarantine zone for suspicious software, allowing you to safely test it without risking damage to your computer.

real-world examples: Used for malware analysis, software testing, and web browsing.

related terms: Virtual Machine, Emulation, Malware Analysis

Sandbox Evasion

definition: Techniques used by malware to detect and avoid being analyzed in a sandbox environment.

explanation: It's like a chameleon changing its colors to blend in with its surroundings. Malware uses sandbox evasion techniques to hide its true behavior and avoid detection.

real-world examples: Malware might check for the presence of a virtual machine or look for specific system characteristics that indicate it's running in a sandbox.

related terms: Malware, Antivirus Evasion, Virtual Machine

Sarbanes-Oxley Act (SOX)

definition: A U.S. federal law that aims to protect investors by improving the accuracy and reliability of corporate disclosures.

explanation: It's like a trust-building measure ensuring companies report their financials honestly and accurately.

real-world examples: Public companies implement internal controls to comply with SOX requirements.

related terms: Financial Reporting, Regulatory Compliance, Corporate Governance

SCADA (Supervisory Control and Data Acquisition)

definition: A system used to monitor and control industrial processes and infrastructure, such as power plants, water treatment facilities, and manufacturing.

explanation: It's like a high-tech control room that oversees and manages critical infrastructure operations.

real-world examples: Utility companies use SCADA systems to monitor and control the flow of electricity across the power grid.

related terms: Industrial Control Systems (ICS), Automation, Remote Monitoring

Scalability

definition: The ability of a system, network, or process to handle a growing amount of work in a capable manner or its ability to be enlarged to accommodate that growth.

explanation: It's like a rubber band that can stretch to accommodate more data or users without breaking.

real-world examples: Cloud computing services are designed to be highly scalable, allowing you to add or remove resources as needed.

related terms: Cloud Computing, Elasticity, Load Balancing

Scanning

definition: The process of examining a system, network, or file for vulnerabilities or threats.

explanation: It's like a security guard checking a building for signs of forced entry.

real-world examples: Port scanning, vulnerability scanning, and malware scanning.

related terms: Vulnerability Assessment, Penetration Testing, Security Assessment

Scareware

definition: A type of malware that uses fear tactics to trick users into buying or downloading unnecessary or harmful software.

explanation: Imagine a pop-up message on your computer warning you of a virus and urging you to click a link to remove it. That link could lead to malware or a fake antivirus program.

real-world examples: Pop-up warnings about viruses or security threats, fake antivirus scans, and messages claiming your computer is infected.

related terms: Malware, Social Engineering, Deception

Screen Scraping

definition: A technique for extracting data from websites by automating the process of copying text and images.

explanation: It's like taking a picture of a webpage and then extracting the data you need from the image.

real-world examples: Used for price comparison websites, data aggregation, and other purposes.

related terms: Web Scraping, Data Extraction, Web Automation

Script Kiddie

definition: An unskilled or inexperienced hacker who uses pre-made tools and scripts to launch attacks.

explanation: It's like a kid playing with a toy gun - they might be able to cause some damage, but they don't really understand how it works.

real-world examples: Script kiddies often deface websites, launch denial-of-service attacks, or spread malware.

related terms: Hacker, Black Hat Hacker, Cybercrime

Secure Access Service Edge (SASE)

definition: A security framework, which is based on a cloud, that combines network security functions with WAN capabilities to support the dynamic secure access needs of organizations.

explanation: It's like a security guard who follows you wherever you go, protecting you from threats no matter where you are.

real-world examples: SASE provides secure access to applications and data from any location, using technologies like SD-WAN and cloud-based security services.

related terms: Cloud Security, SD-WAN, Zero Trust Network Access (ZTNA)

Secure Coding

definition: The practice of developing software with security in mind from the very beginning.

explanation: It's like building a house with strong foundations and reinforced walls to resist earthquakes.

real-world examples: Secure coding practices help prevent vulnerabilities from being introduced into software in the first place.

related terms: Software Development, Application Security, Vulnerability Prevention

Secure Copy Protocol (SCP)

definition: A network protocol used for securely transferring files between a local host and a remote host or between two remote hosts.

explanation: It's like a secure courier service for your files, encrypting them during transit to protect them from prying eyes.

real-world examples: Used to transfer sensitive files, such as confidential documents or financial data.

related terms: SSH, File Transfer, Encryption

Secure Electronic Transaction (SET)

definition: A protocol for securing credit card transactions over the internet.

explanation: It's like a secure tunnel for your credit card information, protecting it from being stolen as it travels across the internet.

real-world examples: SET has largely been replaced by 3D Secure.

related terms: E-commerce, Credit Card Security, Encryption

Secure Enclave

definition: A security coprocessor on Apple devices that provides an isolated environment for storing and processing sensitive data, such as biometric information and encryption keys.

explanation: It's like a separate, heavily guarded room in your house where you keep your most valuable possessions.

real-world examples: Used to protect data from unauthorized access, even if the main processor is compromised.

related terms: Hardware Security, Encryption, Biometrics

Secure File Transfer Protocol (SFTP)

definition: A network protocol for securely transferring files over a network.

explanation: It's like sending a file through a secure tunnel, protecting it from eavesdropping and tampering.

real-world examples: Used for transferring sensitive files, such as financial data or personal information.

related terms: FTP, SSH, Encryption

Secure Hash Algorithm (SHA)

definition: A family of cryptographic hash functions that the National Institute of Standards and Technology (NIST) published.

explanation: It's like a fingerprint for data - it creates a unique, fixed-length string of characters that represents the original data.

real-world examples: Used to verify the integrity of files, passwords, and digital signatures.

related terms: Hash Function, Checksum, SHA-256

Secure Multipurpose Internet Mail Extensions (S/MIME)

definition: A standard for encrypting and digitally signing emails.

explanation: Like sending a letter in a sealed envelope with a tamper-evident seal, S/MIME ensures your emails remain private and unaltered.

real-world examples: Used to protect sensitive emails containing confidential information or financial data.

related terms: Email Encryption, Digital Signature, Public Key Infrastructure (PKI)

Secure Real-time Transport Protocol (SRTP)

definition: A secure version of the Real-time Transport Protocol (RTP) used for transmitting audio and video over IP networks.

explanation: It's like a private, encrypted phone line for your voice and video calls, ensuring confidentiality and preventing eavesdropping.

real-world examples: Used in VoIP applications, video conferencing, and other real-time communication tools to protect against eavesdropping and tampering.

related terms: VoIP, Video Conferencing, Encryption

Secure Shell (SSH)

definition: A cryptographic network protocol for operating network services securely over an unsecured network.

explanation: Imagine a secure tunnel through which you can remotely access and manage a computer system without anyone else being able to see what you're doing.

real-world examples: Used for remote administration, file transfers, and secure remote logins.

related terms: Remote Access, Encryption, SCP (Secure Copy Protocol)

Secure Socket Tunneling Protocol (SSTP)

definition: A form of virtual private network (VPN) tunnel providing a mechanism to transport PPP or L2TP traffic through an SSL/TLS channel.

explanation: It's like a secret passageway through the internet that disguises your traffic as regular web browsing, allowing you to securely access resources on a private network from a remote location.

real-world examples: Used to connect to corporate networks from home or while traveling.

related terms: VPN, SSL/TLS, Remote Access

Secure Sockets Layer (SSL)

definition: A cryptographic protocol designed to provide communications security over a computer network.

explanation: It's the predecessor to TLS and serves the same purpose, ensuring secure communication between web browsers and servers.

real-world examples: While still in use, SSL has been superseded by TLS and is considered less secure.

related terms: TLS, Encryption, HTTPS

Secure Software Development Lifecycle (SDLC)

definition: A process that integrates security at every phase of the software development lifecycle.

explanation: It's like building a house with security features included at every step.

real-world examples: Development teams adopting secure coding practices to prevent vulnerabilities.

related terms: Application Security, Software Development, Cybersecurity

Secure Web Gateway (SWG)

definition: A security solution that filters internet traffic to protect users from web-based threats.

explanation: Think of it as a bouncer for the internet, checking all incoming and outgoing traffic to block malicious content and prevent data leaks.

real-world examples: Used to block access to malicious websites, filter out inappropriate content, and prevent users from downloading malware.

related terms: Web Filtering, Content Filtering, Data Loss Prevention (DLP)

Security Assertion Markup Language (SAML)

definition: An open standard for exchanging authentication and authorization data between parties, specifically, between an identity provider and a service provider.

explanation: It's like a universal passport that allows you to access multiple online services with a single login.

real-world examples: Commonly used for single sign-on (SSO) solutions, where you can use one set of credentials to access multiple applications.

related terms: Single Sign-On (SSO), Identity Management, Authentication

Security Assessment

definition: A systematic evaluation of an organization's security posture to identify vulnerabilities and risks.

explanation: It's like a health checkup for your company's cybersecurity – it assesses the overall health of your security program and identifies areas that need improvement.

real-world examples: Involves vulnerability scans, penetration tests, and reviews of security policies and procedures.

related terms: Vulnerability Assessment, Penetration Testing, Security Audit

Security Audit

definition: An independent examination of an organization's security practices to ensure compliance with regulations and standards.

explanation: It's like a financial audit, but instead of checking financial records, it checks security controls.

real-world examples: Evaluating compliance with regulations like HIPAA or PCI DSS or assessing adherence to industry standards like ISO 27001.

related terms: Compliance, Risk Management, Security Assessment

Security Automation

definition: The use of technology to automate repetitive security tasks, such as vulnerability scanning, patching, and incident response.

explanation: Imagine a robot security guard that tirelessly patrols your network, looking for threats and taking action to neutralize them.

real-world examples: Automated vulnerability scanning tools, security orchestration, automation and response (SOAR) platforms, and robotic process automation (RPA) for security tasks.

related terms: Cybersecurity, Automation, Security Operations

Security Awareness

definition: The understanding of security risks and the knowledge of security best practices.

explanation: It's like knowing how to lock your doors and windows to protect your home from burglars.

real-world examples: Knowing how to create strong passwords, avoid phishing scams, and keep your software up to date.

related terms: Security Awareness Training, Cybersecurity Education, Social Engineering

Security Awareness Training

definition: The process of educating employees about cybersecurity risks and best practices to help them make informed decisions and avoid risky behaviors.

explanation: It's like teaching your employees how to be their own security guards, so they can recognize and respond to threats effectively.

real-world examples: Conducting regular training sessions on topics like phishing, password security, and social engineering.

related terms: Security Awareness, Cybersecurity Education, Employee Training

Security Baseline

definition: A set of minimum security requirements that must be met for a system or network.

explanation: It's like a security checklist—it outlines the essential controls that need to be in place to protect a system.

real-world examples: The Center for Internet Security (CIS) publishes security benchmarks for various operating systems and applications.

related terms: Security Configuration, Hardening, Compliance

Security Benchmark

definition: A set of security standards or best practices that an organization can use to measure its own security posture.

explanation: It's like a yardstick for security - it allows you to compare your organization's security practices to industry standards.

real-world examples: CIS Benchmarks, NIST Cybersecurity Framework, and other security frameworks can be used as benchmarks.

related terms: Security Assessment, Security Baseline, Security Metrics

Security Breach

definition: An incident that results in the unauthorized access, disclosure, alteration, or destruction of sensitive data.

explanation: It's like a hole in a fence that allows an intruder to enter your property and steal your valuables.

real-world examples: Data breaches can result in financial losses, reputational damage, and legal liabilities.

related terms: Data Breach, Cyber Attack, Incident

Security Bulletin

definition: A notification issued by a software vendor or security organization that describes a security vulnerability and recommends how to fix it.

explanation: It's like a warning sign that tells you about a dangerous pothole in the road so you can avoid it.

real-world examples: Microsoft's Security Bulletins, Cisco Security Advisories.

related terms: Vulnerability, Patch, Security Advisory

Security Certification

definition: A credential that demonstrates an individual's knowledge and skills in cybersecurity.

explanation: It's like a diploma or license that proves you have the expertise to do a job.

real-world examples: CISSP, CISA, CISM, CompTIA Security+.

related terms: Cybersecurity, Information Security, Professional Development

Security Clearance

definition: An official authorization granting a person access to classified information or restricted areas.

explanation: It's like a backstage pass that allows you access to areas that are off-limits to the general public.

real-world examples: Government employees and contractors often need security clearances to access classified information.

related terms: National Security, Classified Information, Background Check

Security Compliance

definition: The process of ensuring that an organization's security practices meet regulatory and industry standards.

explanation: It's like making sure your car passes inspection so you can legally drive it on the road.

real-world examples: HIPAA, PCI DSS, GDPR, and other regulations have specific security requirements that organizations must comply with.

related terms: Compliance, Regulatory Compliance, Audit

Security Configuration Management (SCM)

definition: The process of managing and controlling changes to the configuration of systems and networks.

explanation: It's like keeping track of all the changes made to a recipe, so you can reproduce the same results every time.

real-world examples: Used to ensure that systems are configured securely and that changes are made in a controlled and documented manner.

related terms: Configuration Management, Change Management, Security

Security Consulting

definition: The practice of providing expert advice and guidance on cybersecurity matters to organizations.

explanation: It's like hiring a personal trainer for your company's cybersecurity – they assess your strengths and weaknesses, and help you develop a plan to get in shape.

real-world examples: Helping a company implement a new security framework, conducting a risk assessment, or providing training on cybersecurity best practices.

related terms: Cybersecurity, Risk Management, Compliance

Security Control

definition: A safeguard or countermeasure that is designed to protect an organization's assets.

explanation: It's like a lock on a door or a security camera in a store - it's a measure put in place to prevent unauthorized access or theft.

real-world examples: Firewalls, intrusion detection systems, antivirus software, and access controls are all examples of security controls.

related terms: Cybersecurity, Risk Management, Defense in Depth

Security Convergence

definition: The integration of physical security and information security systems and processes.

explanation: It's like combining the police force and the IT department - they work together to protect both physical and digital assets.

real-world examples: Using a single system to manage access control for both buildings and computer networks or integrating video surveillance with intrusion detection systems.

related terms: Physical Security, Information Security, Cybersecurity

Security Engineering

definition: The discipline of designing, building, and implementing secure systems and software.

explanation: It's like an architect designing a building to be earthquake-resistant - security engineers build systems with security in mind from the ground up.

real-world examples: Designing secure software architectures, implementing cryptography, and developing security testing methodologies.

related terms: Cybersecurity, Software Engineering, Systems Engineering

Security Event

definition: Any observable occurrence in a system or network, like a login attempt or file download.

explanation: It's like a blip on a radar screen - it could be anything from a normal login attempt to a malicious attack.

real-world examples: A failed login attempt, a file download, or a change to a system configuration file.

related terms: Security Incident, Log, Event Log Management

Security Event Log

definition: A record of security events that occur on a system or network.

explanation: It's like a security camera's video recording – it captures everything that happens, so you can review it later if something goes wrong.

real-world examples: Windows Event Viewer, Syslog, and application-specific logs.

related terms: Log Management, Security Information and Event Management (SIEM), Incident Response

Security Framework

definition: A set of standards, guidelines, and best practices for managing cybersecurity risk.

explanation: It's like a blueprint for building a secure house, outlining the materials, construction techniques, and safety features needed to keep the occupants safe.

real-world examples: NIST Cybersecurity Framework, ISO 27001, and CIS Controls.

related terms: Cybersecurity, Risk Management, Compliance

Security Governance

definition: The system by which an organization directs and controls its overall security activities and policies.

explanation: It's like the rules and guidelines that ensure everyone follows the same security protocols.

real-world examples: Establishing a security governance framework that includes policies, procedures, and oversight for managing cybersecurity risks.

related terms: Governance, Risk Management, Compliance

Security Hardening

definition: The process of reducing a system's vulnerability by eliminating potential attack vectors and minimizing its attack surface.

explanation: It's like reinforcing the walls of a castle to make it more difficult to breach.

real-world examples: Disabling unnecessary services, applying security patches, and configuring firewalls are all examples of security hardening techniques.

related terms: Vulnerability Management, Patch Management, Configuration Management

Security Incident

definition: A significant event that compromises the security of a system or network, such as a data breach or malware infection.

explanation: It's like a fire alarm going off in the digital world, signaling that something is wrong.

real-world examples: Data breaches, malware infections, denial-of-service attacks, and unauthorized access attempts.

related terms: Cybersecurity, Incident Response, Security Event

Security Incident Response

definition: The process of managing and responding to security incidents.

explanation: It's like a fire drill for cyberattacks – you practice how to respond to an incident so you can act quickly and effectively if one occurs.

real-world examples: Identifying the cause of the incident, containing the damage, eradicating the threat, and recovering normal operations.

related terms: Incident Response Plan, Computer Security Incident Response Team (CSIRT), Security Operations Center (SOC)

Security Information and Event Management (SIEM)

definition: A software solution that aggregates and analyzes security data from multiple sources to identify threats and vulnerabilities.

explanation: It's like a central nervous system for cybersecurity, collecting information from all parts of the body (your network) and alerting the brain (your security team) to potential problems.

real-world examples: SIEM systems can correlate logs from different devices to detect complex attack patterns and provide real-time alerts.

related terms: Log Management, Security Monitoring, Threat Detection

Security Kernel

definition: The central part of a computer's operating system that implements the fundamental security mechanisms.

explanation: It's like the foundation of a house - it provides the basic structure and support for everything else.

real-world examples: The security kernel enforces access control policies, protects memory, and manages communication between processes.

related terms: Operating System, Access Control, Kernel Mode

Security Labeling

definition: The process of assigning security labels to data or resources to indicate their sensitivity level and access restrictions.

explanation: It's like putting different colored labels on files in a filing cabinet, so you know which ones are confidential and which ones can be shared publicly.

real-world examples: Government classification systems (Top Secret, Secret, Confidential) and commercial data classification schemes.

related terms: Data Classification, Access Control, Data Security

Security Misconfiguration

definition: A vulnerability that arises when a system or application is not configured properly.

explanation: It's like leaving a door unlocked - it creates an easy opportunity for attackers to gain unauthorized access.

real-world examples: Weak passwords, unpatched software, and misconfigured firewalls are all examples of security misconfigurations.

related terms: Vulnerability, Configuration Management, Security Hardening

Security Models

definition: Frameworks or theoretical constructs that describe the policies and mechanisms used to enforce security within an organization.

explanation: It's like blueprints that outline how to structure and enforce security measures.

real-world examples: The Bell-LaPadula model is used to ensure data confidentiality in government systems.

related terms: Security Architecture, Access Control Models, Cybersecurity Frameworks

Security Operations Center (SOC)

definition: A centralized unit that deals with security issues on an organizational and technical level.

explanation: It's like a 24/7 security command center, monitoring for threats, responding to incidents, and coordinating security efforts.

real-world examples: SOC teams use a variety of tools and techniques to detect and respond to security threats, including SIEM systems, threat intelligence platforms, and security orchestration and automation tools.

related terms: Cybersecurity Operations, Incident Response, Threat Management

Security Orchestration

definition: The automation of security tasks and workflows to improve efficiency and effectiveness.

explanation: It's like having a team of robots that can handle repetitive security tasks, freeing up human analysts to focus on more complex threats.

real-world examples: Automating incident response workflows, vulnerability scanning, and patch management.

related terms: Security Automation, Security Orchestration, Automation, and Response (SOAR)

Security Orchestration, Automation, and Response (SOAR)

definition: A set of technologies enabling organizations to collect data about security threats and respond to them in a coordinated and automated way.

explanation: It's like a conductor leading an orchestra of security tools, ensuring that they work together harmoniously to defend against attacks.

real-world examples: SOAR platforms can automate incident response, threat hunting, and vulnerability management tasks.

related terms: Security Automation, Security Orchestration, Incident Response

Security Patch

definition: A software update that fixes a security vulnerability.

explanation: It's like a band-aid for a software bug – it fixes a weakness in the code that could be exploited by attackers.

real-world examples: Software vendors regularly release security patches to address new vulnerabilities.

related terms: Vulnerability, Patch Management, Software Update

Security Policy

definition: A document that outlines an organization's security goals, rules, and procedures.

explanation: It's like a rulebook for employees, explaining how they should handle sensitive data, use company computers, and access the network.

real-world examples: Password policies, acceptable use policies, and incident response procedures.

related terms: Cybersecurity, Compliance, Governance

Security Posture

definition: The overall security strength of an organization, including its policies, procedures, and technologies.

explanation: It's like a report card for cybersecurity, assessing an organization's ability to defend against threats.

real-world examples: A strong security posture involves a layered defense approach, with multiple security controls in place to protect against a variety of threats.

related terms: Cybersecurity, Risk Management, Vulnerability Management

Security Procedures

definition: Specific steps and actions taken to ensure the security of an organization's assets.

explanation: It's like detailed instructions on what to do to keep things safe.

real-world examples: Implementing procedures for secure password creation, regular software updates, and data encryption.

related terms: Security Policies, Security Controls, Standard Operating Procedures (SOP)

Security Scorecard

definition: A tool that gives a score to measure the security of an organization.

explanation: It's like a report card that shows how well your security measures are working.

real-world examples: Using a scorecard to evaluate the security of third-party vendors or to track improvements in your own security over time.

related terms: Security Assessment, Risk Management, Cyber Risk Scoring, Security Metrics

Security Standard

definition: A documented set of technical specifications or requirements for establishing and maintaining a secure environment.

explanation: It's like a recipe for security, providing a standardized set of instructions to follow.

real-world examples: ISO 27001, NIST Cybersecurity Framework, PCI DSS.

related terms: Security Compliance, Security Policy, Best Practices

Security Testing

definition: The process of evaluating a system's security by simulating attacks and identifying vulnerabilities.

explanation: Think of it like a stress test for your security defenses - it reveals weaknesses before attackers can exploit them.

real-world examples: Penetration testing, vulnerability scanning, and security audits.

related terms: Vulnerability Assessment, Penetration Testing, Security Assessment

Security Through Obscurity

definition: A flawed security approach that relies on secrecy or complexity to protect a system.

explanation: It's like hiding your valuables under your bed instead of using a safe.

real-world examples: Relying on proprietary encryption algorithms or hiding the details of a system's architecture.

related terms: Defense in Depth, Security by Design, Open Security

Security Token

definition: A physical or digital device that provides authentication for accessing a network or system.

explanation: It's like a key fob for your computer or a security badge for accessing a building.

real-world examples: Hardware tokens like YubiKeys, software tokens like Google Authenticator, and SMS codes sent to your phone.

related terms: Two-Factor Authentication (2FA), Multi-Factor Authentication (MFA), Authentication

Security Token Service (STS)

definition: A service that issues security tokens to clients.

explanation: It's like a ticket booth that issues tickets to people who want to enter a concert or event.

real-world examples: Used in identity federation and single sign-on (SSO) solutions to provide secure access to multiple services.

related terms: SAML, Identity Federation, Single Sign-On (SSO)

Self-XSS

definition: A type of reflected cross-site scripting (XSS) attack where the victim is tricked into executing malicious JavaScript code in their own browser.

explanation: It's like a magician tricking themselves with their own magic trick. The attacker doesn't directly inject the malicious code but rather tricks the victim into doing it themselves.

real-world examples: A malicious link sent via email or social media that, when clicked, executes malicious JavaScript code in the victim's browser.

related terms: Cross-Site Scripting (XSS), Phishing, Social Engineering

Sender Policy Framework (SPF)

definition: An email authentication method designed to detect forging sender addresses during the delivery of the email.

explanation: Imagine it as a return address verification system for emails. It helps prevent spammers from impersonating legitimate senders.

real-world examples: By publishing SPF records in the DNS, domain owners can specify which mail servers are authorized to send email on their behalf.

related terms: Email Authentication, DKIM, DMARC

Sensitive Data / Information

definition: Information that, if lost, compromised, or disclosed without authorization, could result in significant harm, embarrassment, inconvenience, or unfairness to an individual.

explanation: This includes data like social security numbers, financial information, health records, and trade secrets. Protecting sensitive data is crucial to prevent unauthorized access and breaches.

real-world examples: Personally identifiable information (PII), financial information, medical records, and trade secrets.

related terms: Data Classification, Data Protection, Privacy

Sensitive Personal Information (SPI)

definition: A subset of personal information that, if lost, compromised, or disclosed without authorization, could result in substantial harm, embarrassment, inconvenience, or unfairness to an individual.

explanation: It's a type of sensitive data that is particularly personal and requires extra protection.

real-world examples: Social Security numbers, driver's license numbers, financial account information, and health information.

related terms: Personally Identifiable Information (PII), Data Protection, Privacy

Server-Side Attack

definition: An attack that targets vulnerabilities in the server-side of a web application.

explanation: It's like a burglar trying to break into a bank vault, rather than just stealing money from the ATM.

real-world examples: SQL injection, cross-site scripting (XSS), and remote code execution (RCE) are examples of server-side attacks.

related terms: Web Application Security, Vulnerability, Exploit

Server-Side Request Forgery (SSRF)

definition: A web security vulnerability that allows an attacker to induce the server-side application to make HTTP requests to an arbitrary domain of the attacker's choosing.

explanation: It's like tricking a delivery driver into delivering a package to the wrong address.

real-world examples: An attacker could exploit SSRF to access internal systems, scan for vulnerabilities, or even launch attacks on other systems.

related terms: Web Application Security, Vulnerability, Injection Attack

Service Set Identifier (SSID)

definition: A unique identifier for a wireless network.

explanation: It's like the name of your Wi-Fi network – it's how your devices know which network to connect to.

real-world examples: When you connect to a Wi-Fi network, your device will usually display a list of available SSIDs.

related terms: Wi-Fi, Wireless Network, Wireless Access Point

Session Cloning

definition: Hijacking a user's session by copying and using their session ID to impersonate them.

explanation: It's like copying a friend's key to enter their house without permission.

real-world examples: Attackers stealing session cookies to gain unauthorized access to web applications.

related terms: Session Hijacking, Cybersecurity, Web Security

Session Cookie

definition: A cookie that is used to identify a user during a particular session on a website.

explanation: It's like a temporary name tag that a website gives you when you visit, so it can remember who you are as you browse different pages.

real-world examples: Session cookies are typically deleted when you close your browser.

related terms: Cookie, Authentication, Session Management

Session Fixation

definition: A type of web application vulnerability that allows an attacker to hijack a valid user session.

explanation: It's like a thief tricking a car owner into using a counterfeit key, allowing the thief to take over the car once it's started.

real-world examples: Attackers can exploit session fixation vulnerabilities to impersonate users and gain unauthorized access to their accounts.

related terms: Session Hijacking, Web Application Security, Vulnerability

Session Hijacking

definition: An attack that allows an attacker to take over a user's session on a website or application.

explanation: It's like stealing someone's identity while they're online, allowing the attacker to impersonate them and access their accounts.

real-world examples: Attackers can hijack sessions by stealing session cookies, guessing session IDs, or exploiting vulnerabilities in the session management mechanism.

related terms: Session Fixation, Cross-Site Scripting (XSS), Web Application Security

Session Initiation Protocol (SIP)

definition: A signaling protocol used for initiating, maintaining, modifying and terminating real-time sessions that involve video, voice, messaging and other communications applications and services between two or more endpoints on IP networks.

explanation: It's like a phone operator connecting two people on a call – it establishes and manages communication sessions between devices.

real-world examples: Used in VoIP (Voice over IP) systems, video conferencing, and instant messaging.

related terms: VoIP, Video Conferencing, Real-Time Communication

Session Management

definition: The process of managing user sessions on a website or application.

explanation: It's like a bouncer keeping track of who's inside a club and ensuring that only authorized guests remain.

real-world examples: Session management involves generating and assigning session IDs, tracking user activity, and terminating sessions when they are no longer needed.

related terms: Session Cookie, Session Hijacking, Authentication

Session Prediction

definition: An attack that involves guessing a user's session ID to hijack their session.

explanation: It's like trying to guess a combination lock by randomly trying different numbers.

real-world examples: Attackers can automate this process using tools that rapidly generate and test session IDs.

related terms: Session Hijacking, Session Fixation, Web Application Security

Session Replay

definition: Capturing and replaying a user's session to understand their interactions with a website or application.

explanation: It's like recording a video of someone using a computer and playing it back to analyze their actions.

real-world examples: Websites using session replay tools to improve user experience and security.

related terms: User Experience, Cybersecurity, Data Privacy

Shadow IT

definition: The use of information technology systems, devices, software, applications, and services without explicit IT department approval.

explanation: It's like employees bringing their own lunches to work instead of using the company cafeteria - they're using resources outside of the official IT infrastructure.

real-world examples: Employees using personal cloud storage accounts to store company data or installing unauthorized software on their work computers.

related terms: IT Governance, Risk Management, Cybersecurity

Shared Responsibility Model

definition: A cloud security framework that delineates the security obligations of the cloud provider and the cloud customer.

explanation: Think of it like a shared apartment - the landlord is responsible for building security, while the tenant is responsible for locking their own door.

real-world examples: In AWS, Amazon is responsible for the security of the cloud (infrastructure), while customers are responsible for security in the cloud (data and applications).

related terms: Cloud Security, Cloud Service Provider (CSP), Security Responsibility

Shell Shock

definition: A family of security bugs in the Bash shell, a Unix shell and command language.

explanation: It's like a weak spot in the shell of a snail, making it vulnerable to attack.

real-world examples: The Shellshock vulnerability allowed attackers to execute code on vulnerable systems remotely.

related terms: Vulnerability, Bash, Remote Code Execution (RCE)

Shellcode

definition: A small piece of code used as the payload in the exploitation of a software vulnerability.

explanation: Imagine a tiny robot that, once inside a computer, starts wreaking havoc.

real-world examples: Shellcode can be used to open a remote shell, download additional malware, or steal sensitive data.

related terms: Exploit, Payload, Vulnerability

Shimming

definition: A technique used to modify the behavior of a system or application by inserting a layer of code between the original code and the operating system.

explanation: It's like adding a shim to level a wonky table - shimming code modifies the behavior of an application without changing its original code.

real-world examples: Used to fix compatibility issues, add functionality, or bypass security controls.

related terms: API Hooking, DLL Injection, Compatibility

Short Message Service (SMS)

definition: A text messaging service component of most telephone, World Wide Web, and mobile telephony systems.

explanation: It's the technology that allows you to send and receive text messages on your phone.

real-world examples: SMS messages can be used for two-factor authentication (2FA), phishing scams, and other security-related purposes.

related terms: Text Messaging, Two-Factor Authentication, Smishing

Shoulder Surfing

definition: A type of social engineering attack where an attacker observes a victim's screen or keyboard input to steal sensitive information.

explanation: Imagine someone peeking over your shoulder as you enter your PIN at an ATM.

real-world examples: Attackers might try to steal passwords, credit card numbers, or other confidential information by watching you type.

related terms: Social Engineering, Eavesdropping, Privacy

Side Channel Attack

definition: An attack that exploits information leakage from the implementation of a computer system, rather than weaknesses in the implemented algorithm itself.

explanation: It's like a detective analyzing the sounds and vibrations coming from a safe to crack it, rather than trying to pick the lock directly.

real-world examples: Timing attacks, power analysis attacks, and fault injection attacks.

related terms: Cryptanalysis, Encryption, Information Leakage

Sidejacking

definition: The act of intercepting and using an active web session to gain unauthorized access to a user's information.

explanation: It's like eavesdropping on a phone call and joining the conversation without permission.

real-world examples: Attackers using tools to capture session cookies from unencrypted Wi-Fi networks.

related terms: Session Hijacking, Cyberattack, Network Security

Signature-Based Detection

definition: A method of detecting malware by comparing it to a database of known malicious code signatures.

explanation: It's like recognizing a criminal by their fingerprints - if the malware matches a known signature, it's identified as a threat.

real-world examples: Antivirus software typically uses signature-based detection to identify and block known malware.

related terms: Malware Detection, Antivirus, Heuristic Analysis

Silver Ticket Attack

definition: A type of attack that exploits a vulnerability in the Kerberos authentication protocol to create a forged ticket-granting ticket (TGT) for a service.

explanation: It's like a fake backstage pass that gives an attacker access to a specific area of a concert venue.

real-world examples: Attackers can use a silver ticket to impersonate a service and gain unauthorized access to resources on a network.

related terms: Kerberos, Authentication, Privilege Escalation

Simple Mail Transfer Protocol (SMTP)

definition: A communication protocol for electronic mail transmission.

explanation: It's like the postal service for email, responsible for delivering messages from one server to another.

real-world examples: Most email clients and servers use SMTP to send and receive emails.

related terms: Email, Email Server, Mail Transfer Agent (MTA)

Single Sign-On (SSO)

definition: An authentication scheme that allows a user to log in with a single ID and password to any of several related yet independent software systems.

explanation: It's like using one key to unlock multiple doors in a building.

real-world examples: Users can log in once and access multiple applications or websites without having to re-enter their credentials.

related terms: Authentication, Identity Management, Federated Identity

Sinkhole

definition: A DNS server that gives a false result for a domain name or IP address.

explanation: It's like a black hole that absorbs traffic for a specific domain or IP address, preventing it from reaching its intended destination.

real-world examples: Sinkholes can be used to block access to malicious websites or to redirect traffic to a honeypot.

related terms: DNS, DNS Blackhole, Botnet Mitigation

Sinkhole Routing

definition: A network security technique that redirects traffic destined for a malicious or unwanted destination to a sinkhole server.

explanation: It's like redirecting a river to a different course to prevent flooding.

real-world examples: Used to mitigate the impact of attacks or to study attacker behavior.

related terms: DNS Sinkhole, Traffic Redirection, Security Mitigation

Skimming

definition: The act of stealing credit card information using a small device when the card is swiped through a legitimate payment terminal.

explanation: It's like a pickpocket stealing your wallet while you're not looking.

real-world examples: Thieves installing skimmers on ATMs or gas pumps to capture card details.

related terms: Fraud, Data Theft, Physical Security

Smart Card

definition: A plastic card with an embedded microchip that can store and process data.

explanation: It's like a credit card with a brain, capable of storing information and performing calculations.

real-world examples: Used for authentication, secure access, and financial transactions.

related terms: Chip Card, EMV, Two-Factor Authentication (2FA)

Smishing

definition: A type of phishing attack that uses SMS text messages to trick victims into revealing sensitive information.

explanation: It's like a phishing email but delivered via text message instead.

real-world examples: Attackers might send texts that appear to be from your bank, asking you to click on a link or provide your account information.

related terms: Phishing, Social Engineering, SMS

Smurf Attack

definition: A distributed denial-of-service (DDoS) attack flooding a target with Internet Control Message Protocol (ICMP) packets.

explanation: It's like a mob of people sending prank calls to the same phone number, overwhelming the person on the other end.

real-world examples: Attackers use smurf attacks to overload a target's network with traffic, making it unavailable to legitimate users.

related terms: DDoS Attack, ICMP, Amplification Attack

Sniffer

definition: A software or hardware tool that intercepts and analyzes network traffic.

explanation: It's like a wiretap for your network, allowing you to see all the data that is being transmitted.

real-world examples: Used for troubleshooting network problems, monitoring network performance, and detecting security threats.

related terms: Packet Sniffer, Network Monitoring, Traffic Analysis

Snort

definition: An open-source network intrusion detection and prevention system (IDS/IPS).

explanation: It's like a guard dog that alerts you to intruders trying to break into your network.

real-world examples: Organizations using Snort to monitor network traffic for suspicious activity.

related terms: Intrusion Detection System (IDS), Network Security, Cybersecurity

SOC 2

definition: A framework for managing and securing customer data based on five "trust service principles": security, availability, processing integrity, confidentiality, and privacy.

explanation: It's like a seal of approval ensuring service providers meet high standards for data security and privacy.

real-world examples: Cloud service providers obtaining SOC 2 certification to demonstrate their security practices.

related terms: Compliance, Data Security, Privacy

Social Engineering Attack

definition: A type of cyber attack that relies on human interaction to trick victims into revealing sensitive information or performing actions that compromise security.

explanation: Imagine a con artist manipulating someone into giving them money or personal information. Social engineering attacks prey on human vulnerabilities like trust, fear, and greed.

real-world examples: Phishing emails, vishing calls, pretexting, and baiting are common social engineering tactics.

related terms: Phishing, Pretexting, Baiting

Social Engineering Toolkit (SET)

definition: An open-source penetration testing framework designed for social engineering.

explanation: It's like a toolbox filled with tools specifically for tricking and manipulating people.

real-world examples: Using SET to create phishing emails and malicious websites for testing security awareness.

related terms: Social Engineering, Phishing, Penetration Testing

Software as a Service (SaaS)

definition: A software distribution model in which a cloud provider hosts applications and makes them available to customers over the Internet.

explanation: Think of it like renting software instead of buying it. You access the application through your web browser or a mobile app, and the provider takes care of maintenance, updates, and security.

real-world examples: Salesforce, Dropbox, and Google Workspace are examples of SaaS applications.

related terms: Cloud Computing, Cloud Service Provider (CSP), Subscription Model

Software Bill of Materials (SBOM)

definition: A comprehensive list of all components, including software and dependencies, within a piece of software.

explanation: It's like a detailed recipe that lists every ingredient used to make a dish, helping to identify any potential issues.

real-world examples: Using an SBOM to quickly identify and address vulnerabilities in third-party libraries used in an application.

related terms: Software Composition Analysis (SCA), Dependency Management, Vulnerability Management, Software Security

Software Composition Analysis (SCA)

definition: A method of analyzing software to identify open-source and third-party components, as well as any associated security vulnerabilities.

explanation: It's like checking the ingredients list of a food product to see if it contains any allergens or harmful substances. SCA helps organizations identify and address security risks in their software supply chain.

real-world examples: Tools like Black Duck and WhiteSource scan software for known vulnerabilities in open-source components.

related terms: Software Supply Chain, Vulnerability Management, Open Source Software

Software Defined Perimeter (SDP)

definition: A security framework that dynamically creates one-to-one network connections between users and the resources they need to access.

explanation: Imagine a network that only appears when you need it, like a secret door that opens only when you know the right password.

real-world examples: SDP provides secure remote access to applications and data, while reducing the attack surface by hiding network resources from unauthorized users.

related terms: Zero Trust Network Access (ZTNA), Network Security, Access Control

Software Development Life Cycle (SDLC)

definition: A process used by software organizations to design, develop, and test high-quality software.

explanation: It's like a recipe for building software, outlining the steps involved in planning, designing, coding, testing, and deploying an application.

real-world examples: Agile, Waterfall, and DevOps are popular SDLC methodologies.

related terms: Software Development, Software Engineering, Secure Coding

Software Exploitation

definition: The act of taking advantage of vulnerabilities in software in order to gain unauthorized access or control.

explanation: It's like finding a crack in a wall and using it to break into a building.

real-world examples: Exploiting buffer overflow vulnerabilities to execute arbitrary code on a system.

related terms: Vulnerability, Exploit, Malware

Software Vulnerability

definition: A flaw or weakness in a software program that can be exploited by attackers to gain unauthorized access or cause harm.

explanation: It's like a crack in a wall that allows water to seep in - software vulnerabilities are weaknesses that attackers can exploit to compromise a system.

real-world examples: Buffer overflows, SQL injection flaws, and cross-site scripting (XSS) vulnerabilities are common examples.

related terms: Security Bug, Exploit, Patch

Spam

definition: Unsolicited or undesired electronic messages, typically sent in bulk.

explanation: Imagine receiving a flood of unwanted emails or text messages - that's spam.

real-world examples: Spam can be annoying and time-consuming, and it can also contain malicious links or attachments.

related terms: Email Spam, Junk Mail, Phishing

Spam Filter

definition: Software that attempts to block spam messages from reaching a user's inbox.

explanation: It's like a sieve that separates the good emails from the bad ones.

real-world examples: Spam filters use various techniques, such as keyword analysis, Bayesian filtering, and machine learning, to identify and block spam.

related terms: Email Security, Junk Mail, Bayesian Filtering

Spam Trap

definition: An email address that is intentionally created to attract spam.

explanation: It's like a decoy mailbox that's used to collect spam, helping to identify and block spammers.

real-world examples: Spam traps are often used by email providers and anti-spam organizations to improve their filtering capabilities.

related terms: Spam, Email Filtering, Blacklist

Spear Phishing

definition: A highly targeted phishing attack aimed at a specific individual or organization.

explanation: It's like a sniper targeting a specific individual instead of spraying bullets randomly.

real-world examples: Attackers might research their targets to tailor their phishing emails or messages, making them more convincing and difficult to detect.

related terms: Phishing, Social Engineering, Whaling

Spectre & Meltdown Attacks

definition: Hardware vulnerabilities that affect modern processors, allowing attackers to steal sensitive data from memory.

explanation: It's like a superman who can see through walls to steal your secrets. These attacks exploit design flaws in processors to access data that should be protected.

real-world examples: Spectre and Meltdown vulnerabilities were discovered in 2018 and affect a wide range of processors from Intel, AMD, and ARM.

related terms: Hardware Vulnerability, Microarchitecture, Side-Channel Attack

Spoofed Email

definition: An email that appears to have been sent from one source but was actually sent from another source.

explanation: It's like a letter with a fake return address – the sender is trying to trick you into thinking it's from someone else.

real-world examples: Attackers use spoofed emails to impersonate legitimate organizations and trick victims into revealing sensitive information.

related terms: Phishing, Email Spoofing, Spam

Spoofing Attack

definition: A type of attack where an attacker impersonates a trusted entity to gain unauthorized access or information.

explanation: It's like a con artist who uses a fake ID to gain access to a restricted area.

real-world examples: Attackers might use spoofing attacks to bypass authentication mechanisms, steal data, or spread malware.

related terms: Spoofing, Man-in-the-Middle Attack, Phishing

Spyware

definition: Software that secretly gathers information about a person or organization without their knowledge.

explanation: It's like a spy who secretly records your conversations and activities.

real-world examples: Keyloggers, screen capture tools, and tracking cookies are all examples of spyware.

related terms: Malware, Surveillance, Privacy

SQL Injection Attack

definition: An attack that exploits an SQL injection vulnerability.

explanation: This is the act of exploiting the SQL injection vulnerability to gain unauthorized access to or modify data in a database.

real-world examples: Attackers might use SQL injection to steal sensitive information, such as credit card numbers or passwords.

related terms: SQL Injection, Web Application Security, Data Breach

SQL Slammer

definition: A computer worm that caused a denial-of-service attack on some Internet hosts and dramatically slowed down general Internet traffic in 2003.

explanation: This worm exploited a buffer overflow vulnerability in Microsoft SQL Server, spreading rapidly across the internet and causing widespread disruption.

real-world examples: The attack highlighted the importance of patching vulnerabilities and the potential impact of worms on the internet.

related terms: Worm, Malware, DoS Attack

SSL / TSL Certificate

definition: A digital certificate that enables encrypted communication between a web browser and a web server using the SSL/TLS protocol.

explanation: It's like a digital passport for a website, providing proof of its identity and ensuring that communication is encrypted.

real-world examples: When you see the padlock icon in your browser's address bar, it means the website is using an SSL/TLS certificate.

related terms: TLS/SSL, Encryption, Public Key Infrastructure (PKI)

SSL Pinning

definition: A security mechanism that associates a specific cryptographic certificate or public key with a certain website.

explanation: It's like having a photo of your friend's ID stored on your phone so you can verify their identity.

real-world examples: Helps prevent man-in-the-middle attacks by ensuring that the website's certificate matches the one the application expects.

related terms: Man-in-the-Middle Attack, SSL/TLS, Certificate Validation

SSL Stripping

definition: A man-in-the-middle (MitM) attack that downgrades a secure HTTPS connection to an insecure HTTP connection.

explanation: Imagine a thief intercepting a secure package delivery and replacing it with a plain, unsealed box. Attackers use SSL stripping to expose sensitive information that would normally be encrypted.

real-world examples: A user thinks they're logging into their bank's secure website, but the attacker intercepts the traffic and redirects them to an unencrypted version, allowing them to steal login credentials.

related terms: Man-in-the-Middle (MitM) Attack, HTTPS, Encryption

Stack Overflow

definition: A type of buffer overflow that occurs when a program writes more data to a stack buffer than it can hold.

explanation: It's like stacking too many books on a shelf - eventually, the shelf will collapse. In this case, the "collapse" can allow an attacker to execute malicious code.

real-world examples: Attackers can exploit stack overflow vulnerabilities to take control of a program or system.

related terms: Buffer Overflow, Vulnerability, Exploit

Stateful Inspection

definition: A firewall technology that monitors the state of active connections and uses this information to make filtering decisions.

explanation: It's like a doorman who remembers who is allowed into a club and checks their ID every time they enter.

real-world examples: Stateful firewalls can block unauthorized traffic based on the context of a connection, making them more secure than stateless firewalls.

related terms: Firewall, Network Security, Stateless Inspection

Stateless Inspection

definition: A firewall technology that makes filtering decisions based solely on the information in individual packets, without considering the context of a connection.

explanation: It's like a security guard who only checks IDs at the door but doesn't keep track of who is inside the building.

real-world examples: Stateless firewalls are simpler and faster than stateful firewalls, but they are also less secure.

related terms: Firewall, Network Security, Stateful Inspection

State-Sponsored Attack

definition: A cyberattack that is carried out by or on behalf of a nation-state.

explanation: These are attacks launched by governments against other governments or organizations for political, economic, or military purposes.

real-world examples: The Stuxnet worm, which targeted Iranian nuclear facilities, is believed to have been developed by the U.S. and Israel.

related terms: Cyber Warfare, Cyber Espionage, Advanced Persistent Threat (APT)

Static Application Security Testing (SAST)

definition: A method of testing software for vulnerabilities by analyzing its source code without executing it.

explanation: It's like a code inspector who reviews a blueprint before a building is constructed, looking for potential design flaws.

real-world examples: SAST tools can identify common vulnerabilities like SQL injection and buffer overflows.

related terms: Vulnerability Scanning, Security Testing, Code Review

Steganalysis

definition: The practice of detecting and analyzing steganography.

explanation: It's like a detective searching for hidden messages in a seemingly innocent image or audio file.

real-world examples: Steganalysis tools can detect hidden data by analyzing statistical anomalies or using specialized algorithms.

related terms: Steganography, Data Hiding, Covert Channel

Steganography

definition: The practice of concealing messages or information within other non-secret text or data.

explanation: It's like hiding a message in plain sight, such as embedding a secret message in the pixels of an image or the audio samples of a song.

real-world examples: Attackers can use steganography to exfiltrate data or communicate with each other without detection.

related terms: Data Hiding, Covert Channel, Encryption

Stegware

definition: A type of malware that hides malicious code within seemingly harmless files, such as images or videos.

explanation: Imagine a secret message hidden within a picture frame. Stegware works similarly, concealing malicious code within ordinary files to avoid detection.

real-world examples: Attackers might embed malware within an image file attached to an email or hide a malicious script within a video file.

related terms: Malware, Steganography, Data Hiding

Stored XSS

definition: A type of cross-site scripting (XSS) attack where the malicious script is stored on the web server and executed in the victim's browser when they access the affected page.

explanation: It's like leaving a poisoned apple on a table, waiting for someone to take a bite. The malicious code is stored on the website itself and is triggered when the victim visits the page.

real-world examples: A message board where an attacker posts a malicious script that is then executed by anyone who views the message.

related terms: Cross-Site Scripting (XSS), Web Application Security, Injection Attack

Stream Cipher

definition: A type of symmetric encryption algorithm that encrypts data one bit or byte at a time.

explanation: It's like a stream of water flowing through a pipe - the data is encrypted as it flows, and decrypted as it is received.

real-world examples: RC4, ChaCha20, and Salsa20 are examples of stream ciphers.

related terms: Symmetric Encryption, Block Cipher, Cryptography

Strong Password

definition: A password that is hard to guess or crack, typically long, complex, and contains a mix of uppercase and lowercase letters, numbers, and symbols.

explanation: It's like a complex combination lock with multiple dials and numbers - the more complex the password, the harder it is to crack.

real-world examples: A strong password might be something like 3cts3b@tt3ryst@pl3"

related terms: Password Security, Password Complexity, Password Length

Structured Query Language (SQL)

definition: A domain-specific language used in programming and designed for managing data held in a relational database management system (RDBMS), or for stream processing in a relational data stream management system (RDSMS).

explanation: It's the language that computers use to talk to databases, allowing them to store, retrieve, and manipulate data.

real-world examples: SQL is used by a wide range of applications, including web applications, desktop applications, and mobile apps.

related terms: Database, Relational Database Management System (RDBMS), SQL Injection

Stuxnet

definition: A sophisticated computer worm, which targeted industrial control systems, discovered in 2010 that.

explanation: It's like a digital precision weapon designed to sabotage specific industrial processes.

real-world examples: Stuxnet causing physical damage to Iran's nuclear program by altering the speed of centrifuges.

related terms: Cyberweapon, Malware, Industrial Control Systems

Subdomain Takeover

definition: A type of vulnerability that occurs when a subdomain (e.g., [invalid URL removed]) points to a non-existent or expired domain, allowing an attacker to register it and take control of it.

explanation: It's like finding an abandoned house and claiming it as your own.

real-world examples: Attackers can use subdomain takeover to host phishing websites, spread malware, or redirect traffic to malicious websites.

related terms: Domain Name, DNS, Vulnerability

Subnetwork

definition: A logical subdivision of an IP network.

explanation: Think of it like dividing a city into different neighborhoods.

real-world examples: Subnetting allows for more efficient use of IP addresses and can improve network security by isolating different parts of the network.

related terms: IP Address, Subnet Mask, CIDR Notation

Supply Chain Attack

definition: A cyberattack that targets an organization by compromising one of its suppliers or vendors.

explanation: It's like poisoning the water supply upstream – the attacker compromises a supplier's systems to gain access to the target organization's network.

real-world examples: The SolarWinds attack is a recent example of a supply chain attack.

related terms: Third-Party Risk, Vendor Risk Management, Compromise

Supply Chain Risk Management (SCRM)

definition: The process of identifying, assessing, and mitigating risks within the supply chain.

explanation: It's like ensuring all links in the supply chain are strong and secure to prevent disruptions.

real-world examples: Companies assessing suppliers' cybersecurity practices to protect against supply chain attacks.

related terms: Risk Management, Cybersecurity, Supply Chain Security

SWATting

definition: A harassment tactic of deceiving an emergency service (via such means as hoaxing an emergency services dispatcher) into sending a police and emergency service response team to another person's address.

explanation: It's like making a prank call to the police, claiming there's an emergency at someone else's house.

real-world examples: Attackers might target gamers, streamers, or other online personalities, causing a SWAT team to be dispatched to their homes.

related terms: Harassment, Hoax, Emergency Services

Symmetric Encryption

definition: A type of encryption where the same key is used for both encryption and decryption.

explanation: It's like a shared secret between two parties – they both need the same key to unlock the message.

real-world examples: Advanced Encryption Standard (AES), Data Encryption Standard (DES), and Blowfish.

related terms: Encryption, Cryptography, Asymmetric Encryption

SYN Flood Attack

definition: A type of denial-of-service (DoS) attack that overwhelms a server with SYN requests, causing it to exhaust its resources and become unresponsive.

explanation: It's like flooding a restaurant with reservations, preventing legitimate customers from getting a table.

real-world examples: Attackers use SYN floods to disrupt online services, websites, and networks.

related terms: DoS Attack, DDoS Attack, TCP

Syslog

definition: A standard for message logging.

explanation: It's like a centralized logging system for computer networks, collecting log messages from different devices and applications.

real-world examples: Used to monitor system activity, troubleshoot problems, and detect security incidents.

related terms: Log Management, Security Information and Event Management (SIEM), Event Log

System Backdoors

definition: Hidden methods of bypassing normal authentication to gain access to a system.

explanation: It's like a secret door that allows someone to enter a building undetected.

real-world examples: Developers leaving backdoors in software to access systems remotely.

related terms: Malware, Cybersecurity, Exploits

System Hardening

definition: The process of securing a system by reducing its vulnerabilities and attack surface.

explanation: Think of it like reinforcing the doors and windows of your house to make it harder for burglars to break in.

real-world examples: Disabling unnecessary services and ports, applying patches, enforcing strong password policies, and configuring firewalls.

related terms: Security Hardening, Vulnerability Management, Attack Surface Reduction

System Integrity Check

definition: A process that verifies the integrity of a system or file by comparing it to a known good baseline.

explanation: Imagine comparing a photocopy to the original document to see if it has been altered.

real-world examples: File integrity monitoring (FIM) software can detect unauthorized changes to system files and alert administrators to potential security breaches.

related terms: File Integrity Monitoring (FIM), Hash Function, Checksum

System Logging

definition: The process of recording system events and activities in a log file.

explanation: It's like a security camera for your computer, recording everything that happens so you can review it later if something goes wrong.

real-world examples: System logs can be used to detect security incidents, troubleshoot problems, and monitor system performance.

related terms: Log Management, Security Information and Event Management (SIEM), Event Log

T

Tailgating

definition: A physical security breach where an unauthorized person enters a secure area by following an authorized person through a door or other access point.

explanation: It's like sneaking into a concert by following someone with a ticket through the entrance.

real-world examples: An attacker might tailgate an employee through a secure door to gain access to a restricted area.

related terms: Piggybacking, Social Engineering, Physical Security

Targeted Attack

definition: A cyberattack that is specifically aimed at a particular individual, organization, or system.

explanation: It's like a sniper targeting a specific individual instead of shooting randomly into a crowd.

real-world examples: Targeted attacks are often more sophisticated and difficult to defend against than opportunistic attacks.

related terms: Advanced Persistent Threat (APT), Spear Phishing, Cyber Espionage

TCP Hijacking

definition: A type of cyber attack where the attacker takes control of a TCP session between two hosts.

explanation: It's like a hijacker taking over a plane in mid-flight.

real-world examples: Attackers can inject malicious data, eavesdrop on communications, or disrupt the connection entirely.

related terms: Man-in-the-Middle Attack, TCP/IP, Network Security

TCP RST

definition: A TCP packet with the RST (reset) flag set, used to terminate a TCP connection abruptly.

explanation: It's like slamming the phone down on a call.

real-world examples: A TCP RST can be used to close a connection gracefully or to signal an error condition.

related terms: TCP, Network Protocol, Connection Termination

TCP SYN

definition: A TCP packet with the SYN (synchronize) flag set, used to initiate a TCP connection.

explanation: It's like saying "hello" to start a conversation.

real-world examples: The first step in the TCP three-way handshake.

related terms: TCP, Network Protocol, Connection Establishment

Technical Vulnerability

definition: A flaw or weakness in a system's design, implementation, or operation that can be exploited by a threat to affect confidentiality, integrity, or availability adversely.

explanation: It's like a weak spot in a castle wall that an attacker can exploit to gain entry.

real-world examples: Software bugs, misconfigurations, and unpatched vulnerabilities are all examples of technical vulnerabilities.

related terms: Vulnerability, Exploit, Threat

TEMPEST

definition: The study and control of unintentional information leakage via electromagnetic or acoustic signals.

explanation: It's like eavesdropping on a conversation by picking up stray radio waves or sound vibrations.

real-world examples: Shielding sensitive equipment, using TEMPEST-certified devices, and following proper grounding and cabling practices.

related terms: Electromagnetic Interference (EMI), Eavesdropping, Side-Channel Attack

Thin Client

definition: A lightweight computer that relies on a server for processing and storage, typically used in a centralized computing environment.

explanation: It's like a minimalist workspace where all the heavy lifting (processing) is done elsewhere (on the server).

real-world examples: Using thin clients in schools and businesses where users access applications and data stored on a central server.

related terms: Virtual Desktop Infrastructure (VDI), Cloud Computing, Client-Server Model

Third-Party Cookie

definition: A cookie that is set by a domain other than the one you are currently visiting.

explanation: Imagine a salesperson from another store following you around and taking notes on your shopping habits.

real-world examples: Used for tracking, advertising, and other purposes.

related terms: Cookie, First-Party Cookie, Web Tracking

Third-Party Risk Management (TPRM)

definition: The process of identifying, assessing, and controlling risks posed by third-party vendors and service providers.

explanation: It's like making sure the people you hire to help your business don't bring in any hidden dangers.

real-world examples: Evaluating a cloud service provider's security practices before using their services or monitoring a supplier to ensure they comply with data protection regulations.

related terms: Vendor Risk Management, Supply Chain Risk Management, Compliance, Risk Assessment

Threat

definition: Anything that has the potential to cause harm to a system or organization.

explanation: It's like a storm cloud on the horizon - it might not rain, but it's a potential danger.

real-world examples: Natural disasters, hackers, malware, and even disgruntled employees.

related terms: Risk, Vulnerability, Threat Actor

Threat Actor

definition: An individual or group that is responsible for a cyber threat or attack.

explanation: They are the perpetrators of cybercrime, ranging from script kiddies to sophisticated nation-state actors.

real-world examples: Hackers, hacktivists, cybercriminals, and state-sponsored actors.

related terms: Threat, Cyber Attack, Cybercrime

Threat Actor Attribution

definition: The process of identifying the specific threat actor responsible for a cyberattack.

explanation: It's like a detective investigation but in the digital world. The goal is to determine who is behind the attack and what their motives are.

real-world examples: Analyzing malware code, tracking IP addresses, and examining attack patterns.

related terms: Cyber Threat Intelligence (CTI), Incident Response, Digital Forensics

Threat Assessment

definition: The process of identifying and evaluating potential threats to an organization's security.

explanation: It's like scanning the horizon for incoming storms to prepare accordingly.

real-world examples: Conducting regular threat assessments to update security measures.

related terms: Risk Management, Cybersecurity, Incident Response

Threat Emulation

definition: The use of software or tools to simulate real-world cyberattacks.

explanation: It's like a fire drill for your network - you practice responding to a simulated attack to prepare for the real thing.

real-world examples: Used to test the effectiveness of security controls, train security personnel, and identify vulnerabilities.

related terms: Red Teaming, Penetration Testing, Security Exercise

Threat Hunting

definition: The proactive process of searching for and identifying threats that have evaded traditional security defenses.

explanation: It's like a detective searching for clues to a crime that hasn't been reported yet.

real-world examples: Using threat intelligence, behavioral analytics, and other tools to detect signs of compromise that might indicate an ongoing attack.

related terms: Threat Intelligence, Incident Response, Security Operations Center (SOC)

Threat Intelligence

definition: Evidence-based knowledge, including context, mechanisms, indicators, implications and actionable advice, about an existing or emerging menace or hazard to assets that can be used to inform decisions regarding the subject's response to that menace or hazard.

explanation: It's like a weather forecast for cyberattacks, providing information about potential threats so organizations can prepare and defend themselves.

real-world examples: Threat intelligence can come from a variety of sources, including open-source intelligence (OSINT), commercial threat feeds, and government reports.

related terms: Threat Intelligence Platform (TIP), Cybersecurity, Risk Management

Threat Intelligence Platform (TIP)

definition: A software solution that aggregates and analyzes threat intelligence data from multiple sources.

explanation: It's like a news aggregator for cyber threats, collecting and organizing information from various sources to give you a comprehensive view of the threat landscape.

real-world examples: TIPs can help organizations identify and prioritize threats, develop mitigation strategies, and improve their overall security posture.

related terms: Threat Intelligence, Security Information and Event Management (SIEM), Threat Management

Threat Intelligence Sharing

definition: The practice of exchanging information about cyber threats between organizations to improve security.

explanation: It's like a neighborhood watch program where neighbors share information about potential dangers.

real-world examples: Organizations participating in threat intelligence sharing platforms to stay informed about new threats.

related terms: Cybersecurity, Threat Intelligence, Collaboration

Threat Landscape

definition: The overall picture of the cyber threats that an organization faces.

explanation: It's like a map of a battlefield, showing where the enemy is located and what their capabilities are.

real-world examples: The threat landscape is constantly changing as new threats emerge and old ones evolve.

related terms: Threat Intelligence, Risk Assessment, Cybersecurity

Threat Modeling

definition: A process for identifying, understanding, and prioritizing potential threats to a system or application.

explanation: Imagine a security consultant drawing a map of all the ways someone could break into a bank - that's threat modeling. Organizations can prioritize their security efforts and allocate resources more effectively by identifying and evaluating potential threats.

real-world examples: Identifying potential attack vectors, vulnerabilities, and threat actors for a web application.

related terms: Risk Assessment, Attack Surface, Vulnerability Assessment

Threat Surface

definition: The total set of potential vulnerabilities that an attacker could exploit to gain unauthorized access to a system.

explanation: Think of it like the total number of windows and doors in a building - each one is a potential entry point for an intruder.

real-world examples: Open ports, unpatched software, and misconfigured settings are all part of the threat surface.

related terms: Attack Surface, Vulnerability, Threat Vector

Threat Vector

definition: The path or means by which an attacker can gain access to a target system or network to deliver a payload or cause damage.

explanation: It's like the route a burglar takes to break into a house - it could be through an unlocked window, a vulnerable back door, or even a phishing email.

real-world examples: Phishing emails, drive-by downloads, and zero-day vulnerabilities are all examples of threat vectors.

related terms: Attack Vector, Exploit, Vulnerability

Three-Way Handshake

definition: A method used by TCP/IP to establish a reliable connection between a client and a server.

explanation: Imagine it as a three-step conversation: the client says "hello," the server responds with "hello, how are you?", and the client replies "I'm fine, thanks." This confirms that both parties are ready to communicate.

real-world examples: Used to establish TCP connections for web browsing, email, and other internet services.

related terms: TCP/IP, TCP, Connection Establishment

Time-Based Blind SQL Injection

definition: A type of SQL injection attack that relies on timing differences in the server's response to determine if a SQL query was successful.

explanation: It's like a blindfolded person trying to find their way out of a maze by feeling the walls and listening for echoes. The attacker sends carefully crafted queries to the database and observes the response time to infer information about the data.

real-world examples: Attackers can use this technique to extract sensitive data from a database, even if they don't have direct access to the results of the query.

related terms: SQL Injection, Blind SQL Injection, Database Security

Time-Based One-Time Password (TOTP)

definition: A one-time password (OTP) that is generated based on the current time and a shared secret key.

explanation: It's like a password that changes every minute, making it much harder for attackers to guess.

real-world examples: Used in two-factor authentication (2FA) systems, such as Google Authenticator.

related terms: One-Time Password (OTP), Two-Factor Authentication (2FA), Authentication

Time-of-Check to Time-of-Use (TOCTOU)

definition: A type of software vulnerability that happens when a program checks the state of a resource before using it, but the state of the resource changes between the check and the use.

explanation: It's like checking to see if a door is locked, then turning around to find that someone has unlocked it while you weren't looking.

real-world examples: An attacker could exploit a TOCTOU vulnerability to gain unauthorized access to a file or resource.

related terms: Race Condition, Software Vulnerability, Exploit

TimThumb Attack

definition: An attack exploiting a vulnerability in the TimThumb script used in WordPress to resize images.

explanation: It's like a burglar exploiting a known weak spot in a window to break into a house.

real-world examples: Attackers uploading malicious scripts through the TimThumb vulnerability to gain control of websites.

related terms: Web Security, Vulnerability, WordPress

Token

definition: A physical or digital object that represents a right or privilege.

explanation: It's like a concert ticket that allows you to enter a venue. In cybersecurity, tokens are used for authentication, authorization, or to represent a value.

real-world examples: Session tokens, authentication tokens, and security tokens are all examples.

related terms: Authentication, Authorization, Security Token

Tokenization

definition: The process of replacing sensitive data with unique identification symbols that retain all the essential information about the data without compromising its security.

explanation: It's like using a code name for a secret agent - the code name identifies the agent but doesn't reveal their true identity.

real-world examples: Used to protect credit card numbers, social security numbers, and other sensitive data.

related terms: Data Masking, Encryption, Data Security

Traffic Analysis

definition: The process of intercepting and examining messages to deduce information from patterns in communication.

explanation: It's like a detective analyzing traffic patterns to track a suspect's movements.

real-world examples: Used to identify communication patterns, detect anomalies, and gather intelligence.

related terms: Network Monitoring, Security Information and Event Management (SIEM), Eavesdropping

Traffic Interception

definition: Capturing and analyzing data as it is transmitted over a network.

explanation: It's like eavesdropping on a conversation happening over a telephone line.

real-world examples: Using packet sniffers to monitor and analyze network traffic.

related terms: Network Security, Man-in-the-Middle (MitM), Sniffing

Traffic Shaping

definition: A bandwidth management technique that prioritizes certain types of network traffic over others.

explanation: Imagine a highway with multiple lanes - traffic shaping is like assigning different lanes to different types of vehicles (e.g., emergency vehicles get priority over regular traffic).

real-world examples: Used to ensure that critical applications have enough bandwidth, even during periods of high network congestion.

related terms: Quality of Service (QoS), Bandwidth Management, Network Traffic Management

Transmission Control Protocol/Internet Protocol (TCP/IP)

definition: The suite of communication protocols used to interconnect network devices on the internet.

explanation: It's like the language that computers use to communicate with each other over the internet.

real-world examples: TCP/IP is the foundation of the internet and is used by almost all network devices.

related terms: Network Protocol, Internet Protocol (IP), Transmission Control Protocol (TCP)

Transport Layer

definition: The fourth layer in the OSI model, responsible for end-to-end communication between applications.

explanation: It's like the delivery service that ensures your package gets from your house to your friend's house.

real-world examples: TCP and UDP are the two main transport layer protocols.

related terms: OSI Model, TCP, UDP

Transport Layer Security (TLS)

definition: A cryptographic protocol designed to provide communications security over a computer network.

explanation: It's like a secure tunnel for your data, encrypting it so that only the intended recipient can read it.

real-world examples: The successor to SSL, TLS is used to secure web browsing (HTTPS), email, and other internet services.

related terms: SSL, Encryption, HTTPS

Transposition Cipher

definition: A method of encryption by which the positions held by units of plaintext (which are commonly characters or groups of characters) are shifted according to a regular system, so that the ciphertext constitutes a permutation of the plaintext.

explanation: Imagine rearranging the letters in a word to create a code.

real-world examples: The rail fence cipher and the route cipher are examples of transposition ciphers.

related terms: Cipher, Encryption, Cryptography

Trapdoor Function

definition: A function that is easy to compute in one direction but difficult to compute in the opposite direction (finding its inverse) without special information.

explanation: Think of it like a one-way street – it's easy to go down but hard to go back up.

real-world examples: Used in public-key cryptography to create digital signatures and secure key exchange.

related terms: Public Key Cryptography, Encryption, RSA

Triple DES (3DES)

definition: A symmetric-key block cipher, which applies the DES cipher algorithm three times to each data block.

explanation: It's like triple-locking a door for extra security.

real-world examples: Used to be a widely used encryption standard but has been largely replaced by AES.

related terms: Encryption, DES, Symmetric Key Algorithm

Trojan / Trojan Horse

definition: A type of malware that disguises itself as a legitimate program to trick users into installing it.

explanation: It's like a poisoned gift – it looks harmless, but it contains a hidden danger.

real-world examples: Trojans can steal data, install other malware, or give attackers remote access to a system.

related terms: Malware, Virus, Worm

Trojan Downloader / Dropper

definition: A type of Trojan malware that downloads and installs (drops) other malicious software on a victim's computer.

explanation: Think of it like a delivery truck for malware - it may seem harmless at first, but its cargo is dangerous.

real-world examples: A Trojan downloader might disguise itself as a legitimate software update or a pirated movie file.

related terms: Trojan, Malware, Downloader, Payload

Trust Model

definition: A framework that defines the relationships and trust boundaries between different components or entities in a system.

explanation: Think of it as a set of rules for who you trust and under what conditions.

real-world examples: Zero Trust is a security model that assumes no trust by default and requires strict verification for every connection.

related terms: Zero Trust, Security Framework, Authentication

Trusted Execution Environment (TEE)

definition: A secure area inside a main processor that guarantees code and data loaded inside to be protected with respect to confidentiality and integrity.

explanation: It's like a safe room inside your house where you keep your most valuable possessions.

real-world examples: Used in mobile devices to store sensitive data like biometric information and payment credentials.

related terms: Secure Enclave, Hardware Security, Trusted Computing

Trusted Platform Module (TPM)

definition: A hardware chip on a computer's motherboard that provides cryptographic functions.

explanation: Think of it as a built-in security guard for your computer, ensuring the integrity of the system and protecting sensitive data.

real-world examples: Used for secure boot, disk encryption, and password protection.

related terms: Hardware Security Module (HSM), Encryption, Secure Boot

Tunnel

definition: A secure, private pathway through a public network (like the internet).

explanation: Imagine a tunnel under a busy city street that allows you to travel from one point to another without being seen.

real-world examples: Virtual Private Networks (VPNs) use tunneling to encrypt traffic and hide your online activity.

related terms: VPN, Encryption, Encapsulation

Two-Factor Authentication (2FA) / Two-Step Verification

definition: An authentication method that requires two different factors to verify a user's identity.

explanation: It's like having two locks on your door - it makes it much harder for someone to break in.

real-world examples: Using a password and a fingerprint scan, or a password and a one-time code sent to your phone.

related terms: Multi-Factor Authentication (MFA), Authentication, Security

Typosquatting

definition: A type of cyber attack that relies on users mistyping a website's address.

explanation: Imagine typing "Amazom.com" instead of "Amazon.com" - typosquatters register these misspelled domains to redirect users to malicious websites.

real-world examples: Used for phishing attacks, malware distribution, and other scams.

related terms: URL Hijacking, Phishing, Social Engineering

Unified Extensible Firmware Interface (UEFI)

definition: A specification defining a software interface between an operating system and platform firmware.

explanation: It's the modern replacement for the BIOS, offering more advanced features and security capabilities.

real-world examples: UEFI is used to boot up most modern computers.

related terms: BIOS, Bootloader, Firmware

Unified Threat Management (UTM)

definition: A security appliance that combines multiple security functions into a single device.

explanation: It's like a Swiss Army knife for network security, providing a variety of tools in one package.

real-world examples: UTM devices typically include a firewall, intrusion detection/prevention system (IDS/IPS), antivirus, and web filtering.

related terms: Network Security, Firewall, IDS/IPS

Uniform Resource Identifier (URI)

definition: A unique sequence of characters that identifies a logical or physical resource.

explanation: It's like a unique address for any resource on the internet, including web pages, images, and videos.

real-world examples: URLs (Uniform Resource Locators) are a common type of URI.

related terms: URL, Web Address, Internet

Uniform Resource Locator (URL)

definition: The address of a resource on the internet.

explanation: It's like a street address for a website, telling your browser where to find it.

real-world examples: https://www.example.com is a URL.

related terms: URI, Web Address, Domain Name

Uninterruptible Power Supply (UPS)

definition: A device that provides battery backup power in the event of electrical failure or when the electrical power drops to an unacceptable voltage level.

explanation: It's like a backup generator for your computer, keeping it running during a power outage.

real-world examples: Used to protect computers, servers, and other electronic devices from power surges and outages.

related terms: Power Outage, Power Protection, Backup Power

Unpatched Vulnerability

definition: A vulnerability in a software or hardware system that has not been fixed by a patch or update.

explanation: It's like a hole in a fence that hasn't been repaired, leaving your property vulnerable to intruders.

real-world examples: Unpatched vulnerabilities are a major target for attackers, as they provide an easy way to compromise systems.

related terms: Vulnerability, Patch, Exploit

Unsolicited Bulk Email (UBE)

definition: Another term for spam email.

explanation: It's unsolicited email sent in bulk to a large number of recipients.

real-world examples: Spam emails can be annoying, time-consuming, and sometimes dangerous.

related terms: Spam, Junk Mail, Email Security

USA PATRIOT Act

definition: A U.S. law enacted in response to the September 11, 2001, terrorist attacks, aimed at enhancing law enforcement investigatory tools.

explanation: It's like boosting security measures to prevent and respond to terrorism and other threats.

real-world examples: Financial institutions implementing measures to detect and prevent money laundering and terrorist financing.

related terms: Regulatory Compliance, Anti-Money Laundering (AML), National Security

User Account Control (UAC)

definition: A security feature in Microsoft Windows that helps prevent unauthorized changes to a computer system.

explanation: It's like a security guard that asks for your permission before allowing a program to make changes to your computer.

real-world examples: UAC prompts users to confirm actions that could potentially harm their system, such as installing software or changing system settings.

related terms: Windows Security, Privilege Escalation, Security

User Account Control (UAC) Bypass

definition: A technique used by attackers to circumvent User Account Control (UAC) protections.

explanation: It's like finding a way to sneak past the security guard without showing your ID.

real-world examples: Attackers can use various techniques to bypass UAC, such as exploiting vulnerabilities in the UAC mechanism or using social engineering to trick users into approving malicious actions.

related terms: UAC, Privilege Escalation, Malware

User Behavior Analytics (UBA)

definition: A cybersecurity process that involves monitoring user behaviors to detect anomalies and potential threats.

explanation: It's like a detective profiling a suspect to predict their next move.

real-world examples: UBA systems analyze user activity logs to identify unusual patterns that might indicate a security breach or insider threat.

related terms: Security Monitoring, Threat Detection, Anomaly Detection

User Datagram Protocol (UDP)

definition: A connectionless communication protocol used for time-sensitive transmissions over IP networks.

explanation: Imagine sending a postcard - you don't need to establish a formal connection with the recipient, you just send it and hope it arrives. UDP prioritizes speed over reliability, making it suitable for applications like video streaming and online gaming.

real-world examples: Live video streaming, online gaming, DNS lookups.

related terms: TCP/IP, Networking Protocol, Connectionless Protocol

User Entity and Behavior Analytics (UEBA)

definition: A cybersecurity process that monitors user behavior to detect anomalies and potential threats.

explanation: It's like a detective profiling a suspect by analyzing their patterns of behavior. UEBA looks for deviations from normal user activity, which could indicate a security breach or insider threat.

real-world examples: Detecting unusual login patterns, data access attempts, or changes in user behavior.

related terms: Security Analytics, User Monitoring, Insider Threat

User Provisioning

definition: The process of creating, modifying, disabling, and deleting user accounts and their associated permissions.

explanation: It's like the HR department managing employee access to the office – they issue keys, assign desks, and revoke access when someone leaves the company.

real-world examples: Creating a new user account for an employee, giving them access to the files and applications they need, and removing their access when they leave the company.

related terms: Identity Management, Access Management, Onboarding/Offboarding

Variant

definition: A modified version of a malware strain.

explanation: It's like a new strain of the flu - it's the same virus but with some slight changes that make it harder to detect or defend against.

real-world examples: A new version of ransomware that uses a different encryption algorithm or demands a higher ransom.

related terms: Malware, Polymorphic Malware, Antivirus Evasion

Vendor Risk Management (VRM)

definition: The process of identifying, assessing, and mitigating risks associated with third-party vendors.

explanation: It's like doing a background check on a potential business partner - you want to make sure they're trustworthy and won't put your company at risk.

real-world examples: Evaluating the security practices of cloud providers, software vendors, and other third-party suppliers.

related terms: Third-Party Risk Management (TPRM), Supply Chain Security, Due Diligence

Virtual Desktop Infrastructure (VDI)

definition: A technology that allows users to access a desktop operating system and applications remotely from any device.

explanation: Think of it like streaming a movie - you're not actually running the software on your device, but rather accessing it from a remote server.

real-world examples: Allows employees to work from home or on the go, while IT can centrally manage and secure the desktop environment.

related terms: Virtualization, Desktop as a Service (DaaS), Remote Access

Virtual Local Area Network (VLAN)

definition: A logical grouping of network devices behaving as if they were on the same LAN, regardless of their physical location.

explanation: Imagine a group of friends who live in different cities but are all part of the same online gaming community - they can communicate and play together as if they were in the same room.

real-world examples: Used to segment a network into smaller, more manageable groups, improving security and performance.

related terms: Network Segmentation, LAN, Switch

Virtual Machine (VM) Escape

definition: A security vulnerability that allows an attacker to break out of a virtual machine and access the host operating system.

explanation: It's like a prisoner escaping from their cell and gaining access to the rest of the prison.

real-world examples: Attackers might exploit vulnerabilities in the hypervisor or guest operating system to escape a VM.

related terms: Virtualization, Hypervisor, Security Vulnerability

Virtual Private Cloud (VPC)

definition: A virtual network that is logically isolated from other virtual networks in the cloud.

explanation: It's like having your own private section in a shared cloud environment.

real-world examples: VPCs provide a secure and isolated environment for running applications and storing data in the cloud.

related terms: Cloud Computing, Network Security, Isolation

Virtual Private Network (VPN)

definition: A technology that creates a secure, encrypted connection over a public network, like the internet.

explanation: Imagine a tunnel through the internet that protects your data from prying eyes.

real-world examples: Used to access restricted resources, bypass censorship, and protect online privacy.

related terms: Encryption, Tunnel, Remote Access

Virus

definition: A type of malicious software that replicates itself and spreads to other systems.

explanation: It's like a biological virus that infects and spreads among people.

real-world examples: Computer viruses that infect files and spread through email attachments.

related terms: Malware, Cybersecurity, Antivirus

Virus Bulletin

definition: An independent testing organization that evaluates the effectiveness of antivirus and anti-malware products.

explanation: It's like Consumer Reports for antivirus software, providing unbiased reviews and test results.

real-world examples: Security professionals and consumers use Virus Bulletin to choose the best antivirus products for their needs.

related terms: Antivirus, Malware, Security Testing

Virus Hoax

definition: A message that warns users of a non-existent virus or malware threat.

explanation: It's like a false alarm that causes unnecessary panic and confusion.

real-world examples: Virus hoaxes can be spread through email, social media, or other online channels.

related terms: Hoax, Misinformation, Social Engineering

Virus Scanner

definition: A software program that scans files and folders for malware.

explanation: It's like a security guard checking for weapons at the entrance to a building.

real-world examples: Antivirus software typically includes a virus scanner that can detect and remove known malware threats.

related terms: Antivirus, Malware Detection, Signature-Based Detection

Virus Signature

definition: A unique pattern of code or data that identifies a specific virus.

explanation: It's like a fingerprint for a virus, allowing antivirus software to recognize and block it.

real-world examples: Antivirus software uses signature databases to identify known malware threats.

related terms: Malware Signature, Antivirus, Signature-Based Detection

Virus Total

definition: A website that analyzes files and URLs to detect malware.

explanation: It's like a second opinion from a doctor - you upload a file or URL to VirusTotal, and it scans it with multiple antivirus engines to see if it's malicious.

real-world examples: Used by security professionals and consumers to check the safety of files and websites.

related terms: Malware Analysis, Antivirus, Online Scanner

Voice over IP (VoIP)

definition: A technology that allows you to make voice calls over the internet.

explanation: It's like a traditional phone call, but instead of using the phone network, it uses the internet.

real-world examples: Skype, Zoom, and Microsoft Teams are examples of VoIP applications.

related terms: Internet Telephony, Voice Communication, IP Telephony

Voice Phishing (Vishing)

definition: A type of phishing attack that uses voice calls to trick victims into revealing sensitive information.

explanation: It's like a phone call from a fake IRS agent threatening you with legal action if you don't pay a bogus tax bill.

real-world examples: Attackers might spoof caller ID to make it appear that the call is coming from a legitimate source.

related terms: Phishing, Social Engineering, Voice Fraud

Vulnerability

definition: A weakness in a computer system or network, which can be exploited by a threat actor.

explanation: It's like a hole in a fence that allows an intruder to enter your property.

real-world examples: Software bugs, misconfigurations, and weak passwords are all examples of vulnerabilities.

related terms: Exploit, Threat, Risk

Vulnerability Assessment

definition: The process of identifying, classifying, and prioritizing vulnerabilities in a system or network.

explanation: Imagine it as a home inspection for your computer systems, checking for cracks in the walls, faulty wiring, and other weaknesses that could be exploited by intruders.

real-world examples: Using automated tools like Nessus or OpenVAS to scan for vulnerabilities.

related terms: Penetration Testing, Risk Assessment, Vulnerability Scanning

Vulnerability Database

definition: A repository of information about known security vulnerabilities in software, hardware, or operating systems.

explanation: It's like a catalog of weaknesses that have been discovered in various technologies, providing details on how they can be exploited and how to fix them.

real-world examples: National Vulnerability Database (NVD), Common Vulnerabilities and Exposures (CVE), Exploit Database.

related terms: Vulnerability, Exploit, Patch

Vulnerability Disclosure

definition: The practice of responsibly disclosing information about a security vulnerability to the affected vendor or the public.

explanation: It's like informing the building owner about a broken lock so they can fix it before someone takes advantage of it.

real-world examples: Ethical hackers and security researchers often follow coordinated vulnerability disclosure processes to give vendors time to fix the vulnerability before it's made public.

related terms: Responsible Disclosure, Ethical Hacking, Vulnerability Management

Vulnerability Life Cycle

definition: The stages a vulnerability goes through from discovery to remediation.

explanation: It's like the life cycle of a butterfly - from egg to caterpillar to butterfly. Vulnerabilities go through stages of discovery, analysis, reporting, fixing, and verification.

real-world examples: Tracking a vulnerability from the time it's discovered to the time it's patched and verified as fixed.

related terms: Vulnerability Management, Patch Management, Exploit

Vulnerability Management

definition: The cyclical practice of identifying, classifying, prioritizing, remediating, and mitigating vulnerabilities.

explanation: It's like a doctor monitoring a patient's health - they regularly check for signs of illness, diagnose any problems, and prescribe treatment.

real-world examples: Using vulnerability scanners to identify weaknesses, prioritizing them based on risk, and applying patches or other mitigations.

related terms: Vulnerability Assessment, Patch Management, Risk Management

Vulnerability Remediation

definition: The process of fixing a vulnerability in a system or application.

explanation: It's like repairing a hole in a fence to prevent intruders from entering.

real-world examples: Applying a patch, updating software, or changing configurations to address a security weakness.

related terms: Vulnerability, Patch Management, Exploit Mitigation

Vulnerability Scanner

definition: A software tool that scans systems and networks for known vulnerabilities.

explanation: It's like a metal detector that scans for hidden weapons.

real-world examples: Nessus, Qualys, and OpenVAS are examples of vulnerability scanners.

related terms: Vulnerability Assessment, Security Scanning, Penetration Testing

WAF Bypass

definition: A technique used by attackers to evade or circumvent a Web Application Firewall (WAF).

explanation: It's like a burglar finding a way to bypass a security alarm.

real-world examples: Attackers might use obfuscation, encoding, or other techniques to hide malicious traffic from the WAF.

related terms: Web Application Firewall (WAF), Web Application Security, Evasion Technique

War Dialing

definition: A technique used by hackers to discover modems connected to phone lines by dialing a range of phone numbers.

explanation: It's like a telemarketer randomly dialing numbers to find potential customers.

real-world examples: War dialing was more common in the early days of hacking, before the widespread adoption of broadband internet.

related terms: Phreaking, Social Engineering, Modem

War Driving

definition: Searching for Wi-Fi networks while driving around in a vehicle.

explanation: It's like a treasure hunt for Wi-Fi networks, using a laptop or smartphone to detect and map out wireless signals.

real-world examples: War driving can be used for legitimate purposes, such as network mapping or troubleshooting, but can also be used by attackers to identify vulnerable networks.

related terms: Wireless Security, Wi-Fi, Wardriving

War Room

definition: A centralized location where a team of experts monitors, responds to, and manages cybersecurity incidents.

explanation: It's like a command center where generals and soldiers strategize and coordinate their response to a battle.

real-world examples: A security operations center (SOC) where cybersecurity professionals manage and mitigate threats in real-time.

related terms: Security Operations Center (SOC), Incident Response, Cybersecurity Command Center

Warm Site

definition: A backup site that has some equipment and data pre-installed but not enough to immediately resume operations.

explanation: Imagine a spare office space with some furniture and basic equipment, but you still need to bring in your computers and data to get up and running.

real-world examples: Used for disaster recovery when a primary site is unavailable, providing a faster recovery time than a cold site.

related terms: Disaster Recovery, Cold Site, Hot Site

Watering Hole Attack

definition: A targeted attack where the attacker compromises a website or other resource that is frequently visited by the intended victims.

explanation: It's like poisoning a watering hole to infect a herd of animals.

real-world examples: Attackers might compromise a website that is popular with a specific industry or group of users to infect them with malware.

related terms: Targeted Attack, Malware, Drive-by Download

Watermarking

definition: The process of embedding a hidden message or identifier in a digital document, image, audio, or video file.

explanation: It's like a secret message hidden in plain sight.

real-world examples: Used to track the distribution of digital content, protect intellectual property, or identify the source of a leak.

related terms: Digital Rights Management (DRM), Steganography, Copyright Protection

Weak Cipher

definition: An encryption algorithm that is considered insecure due to vulnerabilities or short key lengths.

explanation: It's like a flimsy lock that is easily picked by a burglar.

real-world examples: WEP and DES are examples of weak ciphers that are no longer recommended for use.

related terms: Encryption, Cryptography, Algorithm

Weak Password

definition: A password that is easy to guess or crack.

explanation: It's like using "1234" as your PIN - it's not very secure.

real-world examples: Weak passwords are a common target for attackers, as they can be easily guessed or cracked using brute force attacks.

related terms: Password Strength, Password Policy, Password Cracking

Web Application Firewall (WAF)

definition: A firewall that is specifically designed to protect web applications from attacks.

explanation: It's like a bodyguard for your web applications, shielding them from SQL injection, cross-site scripting (XSS), and other web-based attacks.

real-world examples: WAFs can be hardware appliances, software programs, or cloud-based services.

related terms: Firewall, Web Application Security, Intrusion Prevention System (IPS)

Web Application Security

definition: The practice of securing web applications from attacks and vulnerabilities.

explanation: It's like installing security cameras and alarms in your online store to protect it from hackers and thieves.

real-world examples: Input validation, output encoding, authentication, authorization, and session management.

related terms: Cybersecurity, Web Security, Application Security

Web Beacon

definition: A small transparent image (usually 1 pixel x 1 pixel) that is embedded in a web page or email.

explanation: It's like a hidden tracker that reports back to the sender when the image is loaded.

real-world examples: Web beacons are often used to track email opens, website visits, and user behavior.

related terms: Web Tracking, Tracking Pixel, Privacy

Web Browser Security

definition: The practice of protecting web browsers from attacks and vulnerabilities.

explanation: It's like installing antivirus software on your computer – it protects you from malicious websites and downloads.

real-world examples: Keeping your browser up to date, using security extensions, and being aware of phishing scams.

related terms: Cybersecurity, Browser Exploit, Phishing

Web Content Filtering

definition: The practice of blocking access to specific websites or types of content based on predetermined criteria.

explanation: It's like a parental control setting on a TV that prevents children from watching certain channels.

real-world examples: Used by schools, businesses, and parents to protect users from inappropriate or harmful content.

related terms: Content Filtering, Internet Filtering, Censorship

Web Crawler

definition: An automated program browsing the web, usually with the intention of indexing websites for search engines.

explanation: Think of it like a librarian who systematically browses through books to create a catalog.

real-world examples: Googlebot, Bingbot, and other search engine crawlers are web crawlers that index the content of websites.

related terms: Search Engine Optimization (SEO), Indexing, Bot

Web Defacement

definition: An attack that alters the visual appearance of a website.

explanation: Imagine someone vandalizing a website by changing its content or images.

real-world examples: Hackers might deface a website to make a political statement, promote their own agenda, or simply cause disruption.

related terms: Hacking, Vandalism, Cyber Attack

Web Exploit

definition: A piece of code or software that takes advantage of a vulnerability in a web application or browser.

explanation: It's like a skeleton key that can unlock a door that wasn't meant to be opened.

real-world examples: A web exploit might allow an attacker to inject malicious code, steal data, or take control of a website.

related terms: Vulnerability, Web Application Security, Zero-Day

Web Exploitation Framework

definition: A software platform that provides tools and modules for testing and exploiting web applications.

explanation: It's like a toolkit for ethical hackers and penetration testers, containing a variety of tools for identifying and exploiting web vulnerabilities.

real-world examples: Metasploit, Burp Suite, and OWASP ZAP are popular web exploitation frameworks.

related terms: Web Application Security, Penetration Testing, Vulnerability Assessment

Web Filtering

definition: The process of blocking access to certain websites or web content based on predetermined criteria.

explanation: It's like a parental control setting on a computer that prevents children from accessing inappropriate content.

real-world examples: Web filtering can be used to block access to malicious websites, phishing sites, and sites that contain adult content.

related terms: Content Filtering, Internet Filtering, Censorship

Web Proxy

definition: A server that acts as an intermediary between a user and the internet.

explanation: It's like a middleman between you and the websites you visit.

real-world examples: Web proxies can be used to filter content, cache websites, or provide anonymity.

related terms: Proxy Server, Anonymizer, VPN

Web Proxy Auto-Discovery Protocol (WPAD)

definition: A protocol that allows web browsers to locate and use a web proxy server automatically.

explanation: It's like a GPS for web proxies – it helps your browser find the best proxy server to use.

real-world examples: Often used in corporate environments to enforce web filtering and security policies.

related terms: Web Proxy, Proxy Auto-Configuration (PAC) File, Network Configuration

Web Scraping

definition: The process of extracting data from websites.

explanation: It's like copying information from a website into a spreadsheet.

real-world examples: Web scraping can be used for a variety of purposes, such as price comparison, market research, and contact information gathering.

related terms: Data Extraction, Web Crawler, Bot

Web Server Hardening

definition: The process of securing a web server by reducing its vulnerabilities and attack surface.

explanation: It's like reinforcing the doors and windows of your online store to protect it from thieves.

real-world examples: Disabling unnecessary services, applying security patches, and configuring firewalls are all examples of web server hardening techniques.

related terms: Security Hardening, Vulnerability Management, Web Server Security

Web Server Log

definition: A record of events and activities that occur on a web server.

explanation: It's like a security camera for your website, recording every request and response.

real-world examples: Web server logs can be used to troubleshoot problems, analyze traffic patterns, and detect security incidents.

related terms: Log Analysis, Security Information and Event Management (SIEM), Web Analytics

Web Server Security

definition: The practice of protecting web servers from attacks and vulnerabilities.

explanation: It's like installing a security system in your online store to protect it from hackers and thieves.

real-world examples: Implementing firewalls, intrusion detection systems, encryption, and access controls.

related terms: Web Application Security, Cybersecurity, Network Security

Web Services Security

definition: An extension to SOAP to apply security to Web services.

explanation: It's a set of security standards for web services, ensuring that messages are encrypted and authenticated.

real-world examples: Used to protect sensitive data transmitted between web services.

related terms: SOAP, Web Services, Encryption

Web Shell

definition: A web-based interface that allows an attacker to remotely control a compromised web server.

explanation: It's like a backdoor into your website, giving the attacker complete control over it.

real-world examples: Attackers can use web shells to upload malware, steal data, or deface websites.

related terms: Backdoor, Remote Access, Web Application Security

Whaling

definition: A highly targeted phishing attack aimed at senior executives.

explanation: It's like a fisherman using a harpoon to catch a whale, instead of a regular fishing rod.

real-world examples: Whaling attacks often involve social engineering tactics to trick executives into revealing sensitive information or authorizing fraudulent transactions.

related terms: Phishing, Spear Phishing, CEO Fraud

Whistleblower Policy

definition: Policies and procedures for reporting unethical or illegal activities within an organization.

explanation: It's like providing a safe way for employees to report wrongdoings without fear of retaliation.

real-world examples: Implementing hotlines for anonymous reporting of fraud.

related terms: Ethics, Compliance, Governance

White Box Testing

definition: A security testing method where the tester has full knowledge of the internal workings of the system being tested.

explanation: It's like a mechanic inspecting a car with a detailed diagram of its internal components.

real-world examples: White box testing allows for more thorough and comprehensive testing, as the tester can directly analyze the source code and identify vulnerabilities.

related terms: Black Box Testing, Gray Box Testing, Security Testing

White Hat Hacker

definition: An ethical hacker who uses their skills to test and improve the security of systems and networks.

explanation: They are like the "good guys" of the hacking world, using their knowledge to help organizations find and fix vulnerabilities before the bad guys can exploit them.

real-world examples: White hat hackers often work for security companies or as independent consultants.

related terms: Ethical Hacking, Penetration Testing, Security Researcher

Whitelist

definition: A list of entities (e.g., IP addresses, email addresses, or domains) that are known to be safe and trustworthy.

explanation: It's like a VIP guest list - only those on the list are allowed in.

real-world examples: Whitelisting can be used to restrict access to sensitive systems or to allow only trusted emails through a spam filter.

related terms: Blacklist, Access Control, Email Filtering

Wide Area Network (WAN)

definition: A telecommunications network that extends over a large geographical area, connecting multiple smaller networks, such as local area networks (LANs).

explanation: It's like a vast transportation network that connects cities (LANs) across a country or the world.

real-world examples: The internet itself is a WAN, as it connects millions of smaller networks globally.

related terms: Local Area Network (LAN), Metropolitan Area Network (MAN), Internet

Wi-Fi Eavesdropping

definition: Intercepting and capturing data transmitted over a Wi-Fi network.

explanation: It's like listening in on a private conversation through a thin wall.

real-world examples: Hackers capturing login credentials transmitted over unsecured Wi-Fi networks.

related terms: Network Security, Cybersecurity, Data Theft

Wi-Fi Protected Access (WPA)

definition: A security standard for wireless networks that was designed to replace the insecure WEP standard.

explanation: It's like upgrading the lock on your front door to a more secure model.

real-world examples: WPA uses stronger encryption and authentication mechanisms than WEP.

related terms: Wireless Security, Wi-Fi, WEP

Wi-Fi Protected Access II (WPA2)

definition: A security standard for wireless networks that is more secure than WPA.

explanation: It's like upgrading your lock again - WPA2 provides even stronger encryption and authentication than WPA.

real-world examples: WPA2 is the current standard for Wi-Fi security, but it has been superseded by WPA3.

related terms: Wireless Security, Wi-Fi, WPA

Wi-Fi Protected Access III (WPA3)

definition: The latest security protocol for Wi-Fi networks, designed to be more secure than its predecessors WPA2 and WPA.

explanation: Imagine upgrading the locks on your house to the latest high-security model. WPA3 offers enhanced encryption and protection against brute-force attacks, making it harder for hackers to access your Wi-Fi.

real-world examples: Used in modern routers and devices to secure wireless connections.

related terms: Wi-Fi Security, WPA2, Encryption

WiFi Protected Setup (WPS)

definition: A network security standard that tries to make connections between a router and wireless devices faster and easier.

explanation: Think of it like a shortcut for connecting to Wi-Fi, but it can also be a backdoor for hackers if not used carefully.

real-world examples: WPS often involves pushing a button on the router or entering a PIN, but these methods can be vulnerable to brute-force attacks.

related terms: Wi-Fi Security, Security Vulnerability, Brute Force Attack

Wi-Fi Security

definition: The practice of protecting Wi-Fi networks and devices from unauthorized access, eavesdropping, and other threats.

explanation: It's like putting a fence around your Wi-Fi network to keep out unwanted guests.

real-world examples: Includes using strong passwords, encryption protocols like WPA3, and disabling WPS if not needed.

related terms: Wireless Security, Encryption, Authentication

Wildcard Certificate

definition: A public key certificate that can be used to secure multiple subdomains of a domain.

explanation: It's like a master key that can open all the doors in a building instead of needing a separate key for each door.

real-world examples: A single wildcard certificate can be used to secure "*.example .com," covering all subdomains like "mail.example.com" and "shop.example.com".

related terms: SSL/TLS Certificate, Domain Name, Encryption

Windows Management Instrumentation (WMI)

definition: A set of tools in Windows operating systems for managing and monitoring devices and applications.

explanation: Think of it as a remote control for your Windows computer, allowing administrators to perform tasks remotely.

real-world examples: WMI can be used for legitimate purposes, but attackers can also exploit it to gain control of a system or spread malware.

related terms: Windows, System Administration, Remote Management

Wired Equivalent Privacy (WEP)

definition: An outdated and insecure security protocol for Wi-Fi networks.

explanation: It's like a flimsy lock that's easily picked by a burglar. WEP has been largely replaced by more secure protocols like WPA2 and WPA3.

real-world examples: Avoid using WEP if possible, as it offers very weak protection for your Wi-Fi network.

related terms: Wi-Fi Security, WPA, WPA2

Wireless Access Point (WAP)

definition: A networking device that allows Wi-Fi devices to connect to a wired network.

explanation: It's like a bridge between your wired network and your wireless devices, allowing them to communicate with each other.

real-world examples: Home routers and office Wi-Fi access points are examples of WAPs.

related terms: Wi-Fi, Wireless Network, Router

Wireless Intrusion Detection System (WIDS)

definition: A system that monitors wireless network traffic for suspicious activity and alerts administrators to potential threats.

explanation: Think of it as a security camera for your Wi-Fi network, watching for intruders and sounding an alarm if it sees anything unusual.

real-world examples: WIDS can detect unauthorized access points, deauthentication attacks, and other wireless threats.

related terms: Intrusion Detection System (IDS), Wireless Security, Network Security

Wireless Intrusion Prevention System (WIPS)

definition: A system that not only detects but also actively blocks or mitigates wireless threats.

explanation: It's like a security guard who can not only spot intruders but also physically stop them from entering.

real-world examples: WIPS can automatically block devices that violate security policies or that are suspected of being malicious.

related terms: Intrusion Prevention System (IPS), Wireless Security, Network Security

Wireless LAN Controller (WLC)

definition: A device that manages multiple wireless access points (WAPs) in a centralized way.

explanation: It's like a manager who oversees a team of employees, ensuring that they are working together efficiently.

real-world examples: WLCs can be used to configure, monitor, and troubleshoot WAPs, as well as enforce security policies.

related terms: Wireless Access Point (WAP), Wireless Network, Network Management

Wireless Local Area Network (WLAN)

definition: A wireless computer network that links two or more devices using wireless communication to form a local area network (LAN) within a limited area.

explanation: It's like a group of computers that are connected to each other without wires, using radio waves instead.

real-world examples: Home Wi-Fi networks, coffee shop hotspots, and office wireless networks are all examples of WLANs.

related terms: Wi-Fi, Wireless Network, LAN

Wireless Security

definition: The prevention of unauthorized access or damage to computers or data using wireless networks.

explanation: It's like putting a lock on your Wi-Fi network to keep out unwanted guests.

real-world examples: Includes measures like encryption, authentication, and access controls.

related terms: Cybersecurity, Wi-Fi Security, Network Security

Wireless Security Protocol (WSP)

definition: A set of rules that govern how wireless devices communicate securely over a network.

explanation: Think of it like a secret code that only authorized devices can understand.

real-world examples: WEP, WPA, and WPA2 are all examples of wireless security protocols.

related terms: Wireless Security, Encryption, Authentication

Wireshark

definition: A free and open-source packet analyzer.

explanation: It's like a microscope for network traffic, allowing you to see the individual data packets that make up a communication.

real-world examples: Used for network troubleshooting, analysis, software, and communications protocol development, and education.

related terms: Packet Analyzer, Network Monitoring, Protocol Analysis

Wiretapping

definition: The monitoring of telephone and Internet conversations by a third party, often by covert means.

explanation: It's like someone secretly listening in on your phone calls.

real-world examples: Law enforcement agencies may use wiretapping with a warrant to investigate crimes.

related terms: Surveillance, Eavesdropping, Privacy

Worm

definition: A standalone malware computer program that replicates itself in order to spread to other computers.

explanation: Imagine a virus that can spread from one computer to another without any human interaction.

real-world examples: Worms can spread quickly and cause widespread damage by consuming network bandwidth or deleting files.

related terms: Malware, Virus, Trojan

WPA/WPA2 Cracking

definition: The process of breaking the Wi-Fi Protected Access (WPA) or WPA2 security protocols to gain unauthorized access to a wireless network.

explanation: It's like picking the lock on a high-security door to enter a restricted area.

real-world examples: Using tools like Aircrack-ng to crack WPA/WPA2 passwords.

related terms: Wi-Fi Security, Network Security, Wireless Security

X

X.509 Certificate

definition: A digital certificate that uses the X.509 public key infrastructure (PKI) standard to verify the identity of a person, device, or service.

explanation: Imagine it as a digital passport that proves who you are online. It contains information like your name, public key, and the signature of a trusted authority (Certificate Authority).

real-world examples: Used in SSL/TLS for secure web browsing, email encryption, and code signing.

related terms: Digital Certificate, Public Key Infrastructure (PKI), SSL/TLS

Xmas Attack

definition: A type of network scan that sends packets with various TCP flags set to probe for open ports and vulnerabilities.

explanation: It's like a Christmas tree with all the lights on - the attacker sends a bunch of different signals to see which ones get a response.

real-world examples: Used by hackers to discover open ports and identify potential vulnerabilities in a system.

related terms: Port Scanning, Network Scanning, Nmap

XML Bomb

definition: An attack that exploits vulnerabilities in XML parsers to consume excessive resources, potentially crashing the system.

explanation: Imagine a tiny snowball that triggers a massive avalanche. A small XML file can be crafted to consume massive amounts of memory or processing power when parsed, causing a denial of service (DoS) attack.

real-world examples: Billion Laughs attack, Quadratic Blowup attack.

related terms: XML External Entity (XXE), DoS Attack, XML Parsing

XML Encryption

definition: The process of encrypting XML data to protect its confidentiality.

explanation: It's like putting a lock on an XML file so only authorized users can access its contents.

real-world examples: Used to protect sensitive data transmitted in XML format, such as medical records or financial transactions.

related terms: XML, Encryption, Cryptography

XML External Entity (XXE)

definition: A type of attack that exploits vulnerabilities in XML parsers by referencing external entities within an XML document.

explanation: It's like a hacker sneaking a secret message into a letter by referencing a hidden document.

real-world examples: Attackers can use XXE attacks to access sensitive files on a server, execute remote code, or launch denial-of-service (DoS) attacks.

related terms: XML, XML Bomb, Injection Attack

XML Signature

definition: A digital signature for XML documents.

explanation: It's like a handwritten signature on a paper document but for XML files.

real-world examples: Used to verify the authenticity and integrity of XML data.

related terms: Digital Signature, XML, Cryptography

Zero Trust

definition: A security framework that assumes no user or device can be trusted by default, even if they are within the organization's network.

explanation: It's like a bouncer at a club who checks everyone's ID, even if they look familiar.

real-world examples: Zero Trust requires strict verification for every access request, regardless of the user's location or device.

related terms: Network Security, Access Control, Micro-Segmentation

Zero Trust Architecture

definition: A security architecture that implements the Zero Trust principles.

explanation: It's like building a fortress with multiple layers of security, where every door requires a keycard and every room has its own alarm system.

real-world examples: Zero Trust architectures use micro-segmentation, identity, and access management (IAM), and other security controls to create a highly secure environment.

related terms: Zero Trust, Network Security, Microsegmentation

Zero Trust Network Access (ZTNA)

definition: A security framework that provides secure remote access to applications and resources based on the Zero Trust principles.

explanation: Imagine a tunnel that only opens for authorized users, verifying their identity and device security before granting access.

real-world examples: ZTNA solutions use a variety of technologies, such as software-defined perimeters (SDP) and identity-based access control, to provide secure remote access.

related terms: Zero Trust, Secure Access Service Edge (SASE), VPN

Zero-Day Attack / Exploit / Vulnerability

definition: An attack that exploits a vulnerability that is unknown to the software vendor or security researchers.

explanation: It's like a surprise attack - you don't have time to prepare or defend yourself because you didn't know it was coming.

real-world examples: Zero-day attacks are often very effective because there are no patches or mitigations available.

related terms: Vulnerability, Exploit, Patch

Zero-Knowledge Proof

definition: A method by which one party (the prover) can prove to another party (the verifier) that they know a value x, without conveying any information apart from the fact that they know the value x.

explanation: It's like proving you know a secret password without revealing the password.

real-world examples: Used in authentication protocols and privacy-preserving technologies like cryptocurrency.

related terms: Cryptography, Authentication, Privacy

Zombie

definition: A computer that has been compromised by a hacker and is being used to perform malicious activities without the owner's knowledge.

explanation: Imagine a puppet being controlled by a puppeteer. A zombie computer is controlled by a hacker to send spam, participate in DDoS attacks, or steal data.

real-world examples: Botnets are made up of thousands or even millions of zombie computers.

related terms: Botnet, Malware, Command and Control

Full List of Terms

Access Control

Access Management

Access Token

Account Takeover (ATO)

Active Cyber Defense (ACD)

Active Directory

Active Reconnaissance

Adaptive Authentication

Advanced Encryption Standard (AES)

Advanced Persistent Threat (APT)

Advanced Threat Protection (ATP)

Adversarial Machine Learning

Adversary Emulation

Adware

Agile Development

AI-powered attacks

Air Gap

Air-Gapped Network

Algorithm

ANNUALIZED LOSS EXPECTANCY (ALE)

ANNUALIZED RATE OF OCCURENCE (ARO)

ANONYMIZER

ANTI-FORENSICS

ANTI-MONEY LAUNDERING (AML)

ANTI-TAMPER TECHNOLOGY

ANTIVIRUS

API ABUSE

APPLICATION CONTROL

APPLICATION FIREWALL

APPLICATION LAYER

APPLICATION PROGRAMMING INTERFACE (API)

APPLICATION PROGRAMMING INTERFACE (API) GATEWAY

APPLICATION PROGRAMMING INTERFACE SECURITY (API SECURITY)

APPLICATION SECURITY

APPLICATION WHITELISTING

ARMORED VIRUS

ARP SPOOFING (POISONING)

ARTIFICIAL INTELLIGENCE (AI)

ASSET

ASSET INVENTORY

ASSET MANAGEMENT

ASYMMETRIC ENCRYPTION

ASYNCHRONOUS ENCRYPTION

ATM JACKPOTTING

ATTACK ATTRIBUTION

ATTACK GRAPH

ATTACK SURFACE

ATTACK VECTOR

AUDIT EVIDENCE

AUDIT TRAIL

AUTHENTICATION

AUTHENTICATION FACTOR

AUTHENTICATION HEADER (AH)

AUTHORIZATION

AUTOMATED THREAT INTELLIGENCE

AVAILABILITY

AVAILABILITY ZONE

BACKDOOR

BACKSCATTER

BACKTRACE

BAIT AND SWITCH

BAITING

BASEL III

BASELINE

BASTION HOST

BEACONING

BEC (BUSINESS EMAIL COMPROMISE)

BEHAVIOR ANOMALY DETECTION

BEHAVIORAL ANALYTICS

BEHAVIOR-BASED DETECTION / SECURITY

BIOMETRIC AUTHENTICATION

BIOMETRICS

BIRTHDAY ATTACK

BIRTHDAY PARADOX

BLACK BOX TESTING

BLACK HAT HACKER

BLACK HOLE

Business Email Compromise (BEC)

Business Impact Analysis (BIA)

Business Logic Attack

Business Logic Vulnerability

Cache

Cache Poisoning

California Consumer Privacy Act (CCPA)

Call Back

Canary Token

CAPTCHA

CAPTCHA Farms

Card Cloning

Card Skimming

Carding

Certificate Authority (CA)

Certificate Management

Certificate Pinning

Certificate Revocation List (CRL)

Certificate Transparency

Chain of Custody

Chain of Trust

Challenge-Handshake Authentication Protocol (CHAP)

Challenge-Response Authentication

Chatbot Security

Checksum

Chief Information Security Officer (CISO)

Child Sexual Abuse Material (CSAM)

Children's Online Privacy Protection Act (COPPA)

Chopper

Collision Attack

Command and Control (C2)

Command Injection

Common Platform Enumeration (CPE)

Common Vulnerabilities and Exposures (CVE)

Compliance

Compliance Audit

Compliance enforcement

Compliance obligations

Compliance officer

Compliance program

Compliance risk assessment

Compromise Assessment

Computer Emergency Response Team (CERT)

Computer Fraud and Abuse Act (CFAA)

Computer Security

Computer Security Incident Response Team (CSIRT)

Confidentiality

Confidentiality Agreement

Configuration management

Conflict of interest policy

Consumer Privacy Protection Act

Container Security

Containerization

Containment

Content Disarm and Reconstruction (CDR)

Content Filtering

Content Security Policy (CSP)

Continuous Adaptive Risk and Trust Assessment (CARTA)

Control objectives

Control Plane

Control self-assessment (CSA)

Cookie

Cookie Hijacking

Cookie Poisoning

Corporate Espionage

Credential Dumping

Credential Harvesting

Credential Management

Credential reuse

Credential Stuffing

Critical Security Control (CSC)

Cross-Site Request Forgery (CSRF)

Cross-Site Scripting (XSS)

Cross-Site Tracing (XST)

Cryptanalysis

Crypter

Crypto Locker

Crypto-Agility

Cryptocurrency security

Cryptographic Hash Function

Cryptographic Key Management

Cryptographic Nonce

Cryptographic Primitive

Cryptography

Cryptojacking

Crypto-Malware

Cryptomining

CRYPTOMINING MALWARE

CYBER ATTACK

CYBER COMMAND

CYBER DECEPTION

CYBER ESPIONAGE

CYBER HYGIENE

CYBER INCIDENT

CYBER INSURANCE

CYBER KILL CHAIN

CYBER RANGE

CYBER RESILIENCE

CYBER RESILIENCE FRAMEWORK

CYBER SQUAD

CYBER SQUATTING

CYBER THREAT ACTOR (CTA)

CYBER THREAT HUNTING

CYBER THREAT INFORMATION SHARING (CTIS)

CYBER THREAT INTELLIGENCE (CTI)

CYBER VANDALISM

CYBERBULLYING

CYBERCRIME

CYBERCRIME INVESTIGATION

CYBERCRIME-AS-A-SERVICE (CAAS)

CYBERDEFENSE

CYBERESPIONAGE

CYBERSECURITY

CYBERSECURITY AUDIT

CYBERSECURITY AWARENESS MONTH

CYBERSECURITY CAPABILITY MATURITY MODEL (C2M2)

Cybersecurity Education

Cybersecurity Framework

Cybersecurity Insurance

Cybersecurity maturity model

Cybersecurity Maturity Model Certification (CMMC)

Cybersecurity Mesh

Cybersecurity Operations

Cybersecurity posture

Cybersecurity Risk Management Framework (RMF)

Cybersecurity Strategy and Implementation Plan (CSIP)

Cyberterrorism

Cyberwarfare

Cypherpunk

DAD Triad

Dark Web

Data Accountability and Trust Act (DATA)

Data Anonymization

Data at Rest

Data Backups

Data Breach

Data Center Security

Data Centric Security

Data Classification

Data Classification Scheme

Data Controller

Data Custodian

Data Destruction

Data Diddling

Data Diode

DATA DISCOVERY

DATA ENCRYPTION

DATA ENCRYPTION STANDARD (DES)

DATA ERASURE

DATA EXFILTRATION

DATA GOVERNANCE

DATA GOVERNANCE ACT (DGA)

DATA IN TRANSIT

DATA INTEGRITY

DATA INTEGRITY CHECK

DATA INTERCEPTION

DATA LEAK

DATA LEAKAGE

DATA LOSS PREVENTION (DLP)

DATA MASKING

DATA MINIMIZATION

DATA NORMALIZATION

DATA PRIVACY

DATA PRIVACY FRAMEWORK

DATA PROCESSING AGREEMENT (DPA)

DATA PROCESSOR

DATA PROFILING

DATA PROTECTION IMPACT ASSESSMENT (DPIA)

DATA PROTECTION OFFICER (DPO)

DATA RECOVERY

DATA REMANENCE

DATA RETENTION

DATA RETENTION POLICY

DATA SANITIZATION

Data Security

Data Sensitivity

Data Sovereignty

Data Subject

Data Subject Access Request (DSAR)

Data Subject Rights

Data Tampering

Database Activity Monitoring (DAM)

Database Firewall

Database Security

Decentralized Identity

Deception Technology

Decryption

Decryption Oracle

Deep Packet Inspection (DPI)

Deep Web

Deepfake

Defacement

Default Password

Defense Advanced Research Projects Agency (DARPA)

Defense in Depth

Degaussing

Demilitarized Zone (DMZ) Host

Denial of Service (DoS)

Denial of Service Attack

Denial-of-Service as a Service (DDoSaaS)

DevSecOps

Dictionary Attack

Dictionary Wordlist

Differential Cryptanalysis

Digital Certificate

Digital Forensics

Digital Identity Verification

Digital Millennium Copyright Act (DMCA)

Digital Millennium Copyright Act (DMCA) Takedown Notice

Digital Operational Resilience Act (DORA)

Digital Rights Management (DRM)

Digital Risk Protection (DRP)

Digital Signature

Directive on Security of Network and Information Systems (NIS Directive)

Directory Harvest Attack

Directory Services

Directory Traversal Attack

Disaster Recovery (DR)

Disaster Recovery as a Service (DRaaS)

Disaster recovery planning (DRP)

Disk Encryption

Distributed Denial of Service (DDoS)

Distributed reflection denial of service (DRDoS)

DMZ (Demilitarized Zone)

DNS Amplification Attack

DNS Blackhole

DNS Cache

DNS Cache Poisoning

DNS Filtering

DNS Flood

DNS Hijacking

DNS Poisoning

DNS Resolver

DNS Sinkhole

DNS Spoofing

DNS Tunneling

DNS Zone Transfer

DNSSEC

Document Exploitation (DOCEX)

Domain Generation Algorithm (DGA)

Domain Keys Identified Mail (DKIM)

Domain Name Registrar

Domain Name System (DNS)

Domain name system (DNS) security

Domain Privacy Protection

Domain Reputation

Domain Shadowing

Domain Squatting

Dormant Account

DOS Attack

Double Tagging

DoublePulsar

Downgrade Attack

Downstream Liability

Doxing

Doxware

Drive-by Attack

Drive-by Download

Drop Attack

Dropper

Dual-Homed Host

DUE DILIGENCE

DUMPSTER DIVING

DYNAMIC ANALYSIS

DYNAMIC APPLICATION SECURITY TESTING (DAST)

EAVESDROPPING

EGRESS FILTERING

ELLIPTIC CURVE CRYPTOGRAPHY (ECC)

EMAIL SECURITY

EMBEDDED MALWARE

EMBEDDED SYSTEM SECURITY

EMERGENCY MANAGEMENT (RESPONSE)

ENCAPSULATION

ENCAPSULATION SECURITY PAYLOAD (ESP)

ENCRYPTED COMMUNICATION

ENCRYPTED VIRUS

ENCRYPTION

ENCRYPTION ALGORITHM

ENCRYPTION KEY

ENDPOINT

ENDPOINT DETECTION AND RESPONSE (EDR)

ENDPOINT PROTECTION PLATFORM (EPP)

ENDPOINT SECURITY

END-TO-END ENCRYPTION (E2EE)

ENTERPRISE RISK ASSESSMENT

ENTERPRISE RISK MANAGEMENT (ERM)

EPHEMERAL PORT

ESG (ENVIRONMENTAL, SOCIAL, AND GOVERNANCE)

ETHICAL HACKER

ETHICAL HACKING

FAULT INJECTION ATTACK

FEDERATED IDENTITY

FEDERATED IDENTITY MANAGEMENT (FIM)

FILE INCLUSION VULNERABILITY

FILE INTEGRITY MONITORING (FIM)

FILE TRANSFER PROTOCOL (FTP)

FILE TRANSFER PROTOCOL SECURE (FTPS)

FILE VAULT

FILELESS ATTACK

FILELESS MALWARE

FINGERPRINTING

FIREWALKING

FIREWALL

FIREWALL LOG

FIREWALL POLICY

FIREWALL RULE

FIRMWARE

FIRMWARE ANALYSIS

FIRMWARE ATTACKS

FIRST-PARTY COOKIE

FLASH COOKIE

FOOTPRINTING

FORENSIC ANALYSIS

FORENSIC IMAGE

FORENSIC INVESTIGATOR

FORMJACKING

FORWARD SECRECY

FORWARDED EVENTS

FRAUD DETECTION

Fraud risk management

Full Disk Encryption (FDE)

Fuzz Testing / Fuzzing

Gateway

Gateway Firewall

General Data Protection Regulation (GDPR)

Generic Routing Encapsulation (GRE)

Geographic Information System (GIS)

Geolocation Spoofing

Ghostware

Global Positioning System (GPS) Spoofing

Golden Ticket Attack

Governance framework

Governance, Risk, and Compliance (GRC)

Gramm-Leach-Bliley Act (GLBA)

Gray Hat Hacker

Grayware

Green hat hacker

Group Policy

Hacker

Hacker Ethics

Hacker Group

Hacktivist

Hardening

Hardware Firewall

Hardware Keylogger

Hardware Security Module (HSM)

Hash

Hash Collision

HASH FUNCTION

HASHCAT

HASHING

HEADER

HEADER MANIPULATION

HEALTH INSURANCE PORTABILITY AND ACCOUNTABILITY ACT (HIPAA)

HEARTBLEED

HEARTBLEED BUG

HEURISTIC ANALYSIS

HEURISTIC ANTIVIRUS

HIDDEN FIELD

HIDDEN FILE

HIGH ASSURANCE

HIGH AVAILABILITY (HA)

HIJACKING

HIPAA

HONEYNET

HONEYPOT

HOST INTRUSION PREVENTION SYSTEM (HIPS)

HOST-BASED FIREWALL (HBF)

HOST-BASED INTRUSION DETECTION SYSTEM (HIDS)

HOT SITE

HOTFIX

HTML INJECTION

HTTP RESPONSE SPLITTING

HTTP STRICT TRANSPORT SECURITY (HSTS)

HTTPS

HUMAN FIREWALL

HUMAN-COMPUTER INTERACTION (HCI)

Hybrid Analysis

Hybrid Attack

Hybrid Cloud

Hybrid Cloud Security

Hybrid Encryption

Hybrid Warfare

Hyperjacking

Hypertext Preprocessor (PHP)

Hypertext Transfer Protocol (HTTP)

Hypertext Transfer Protocol Secure (HTTPS)

Hypervisor

ICMP Flood

Identity and Access Management (IAM)

Identity as a Service (IDaaS)

Identity Federation

Identity Governance and Administration (IGA)

Identity Management

Identity Proofing

Identity Provider (IdP)

Identity Theft

Image Analysis

Image Spam

Impersonation Attack

Implicit Deny

Incident

Incident Management

Incident Response

Incident Response Plan

Incident Response Team (IRT)

Indicator of Attack (IoA)

Indicators of Compromise (IoCs)

Industrial Control System (ICS)

Industrial Espionage

Industrial IoT attacks

Information Assurance (IA)

Information Leakage

Information Rights Management (IRM)

Information Security

Information security management

Information Security Management System (ISMS)

Information Security Policy

Information sharing and analysis center (ISAC)

Information Warfare

InfoSec

Infrastructure as a Service (IaaS)

Infrastructure as Code (IaC)

Injection Attack

Injection Vulnerability

Input Sanitization

Input Validation

Insecure Deserialization

Insecure Direct Object Reference (IDOR)

Insider

Insider Attack

Insider Threat

Insider Threat Program

Integrated Development Environment (IDE)

Integrity

INTEGRITY MEASUREMENT

INTELLECTUAL PROPERTY (IP) THEFT

INTELLECTUAL PROPERTY RIGHTS (IPR)

INTERCEPTION PROXY

INTERNAL AUDIT

INTERNAL CONTROLS

INTERNAL NETWORK PENETRATION TEST

INTERNAL VULNERABILITY SCAN

INTERNET CONTROL MESSAGE PROTOCOL (ICMP)

INTERNET ENGINEERING TASK FORCE (IETF)

INTERNET MESSAGE ACCESS PROTOCOL (IMAP)

INTERNET OF THINGS (IOT)

INTERNET PROTOCOL (IP)

INTERNET PROTOCOL (IP) CAMERA HACKING

INTERNET PROTOCOL SECURITY (IPSEC)

INTERNET PROTOCOL VERSION 4 (IPV4)

INTERNET PROTOCOL VERSION 6 (IPV6)

INTERNET RELAY CHAT (IRC)

INTERNET STORM CENTER (ISC)

INTRUSION

INTRUSION DETECTION

INTRUSION DETECTION SYSTEM (IDS)

INTRUSION DETECTION SYSTEM EVASION

INTRUSION KILL CHAIN

INTRUSION PREVENTION SYSTEM (IPS)

INTRUSION SIGNATURE

INTRUSION TOLERANCE

INTRUSIVE VULNERABILITY SCAN

IOT ATTACKS

IoT Security

IP Address

IP Address Spoofing

IP Blacklisting

IP Fragmentation

IP Fragmentation Attack

IP Reputation

IP Spoofing

IP Whitelisting

ISO/IEC 27001

IT governance

Jailbreaking

JavaScript Hijacking

JavaScript Object Notation (JSON)

Kerberos

Key Derivation Function (KDF)

Key Escrow

Key Exchange

Key Management System (KMS)

Key performance indicators (KPIs)

Key risk indicators (KRIs)

Key Stretching

Keylogger

Keystroke Dynamics

Kill Chain

Know Your Customer (KYC)

Known Bad

Known Good

Known Plaintext Attack

Malicious Attachment

Malicious Code

Malicious Domain

Malicious Insider

Malicious URL

Malvertising

Malware

Malware Analysis

Malware as a Service (MaaS)

Malware Removal

Malware Sandbox

Malware Signature

Managed Detection and Response (MDR)

Man-in-the-Browser (MitB) Attack

Man-in-the-Cloud Attack

Man-in-the-Middle (MitM) Attack

Masquerade Attack

Master Boot Record (MBR)

Master Service Agreement (MSA)

Mean Time to Detect (MTTD)

Mean Time to Respond (MTTR)

Meet-in-the-Middle Attack

Memory Corruption

Memory Dump

Memory Forensics

Memory Leak

Message Authentication Code (MAC)

Metadata

Metamorphic Malware

Metasploit

Microsegmentation

MITRE ATT&CK

Mobile Application Management (MAM)

Mobile Application Security

Mobile device exploits

Mobile Device Management (MDM)

Mobile Malware

Mobile Security

Mobile Threat Defense (MTD)

Multi-factor Authentication (MFA)

Multipartite Virus

Multi-Party Computation (MPC)

Multi-Tenancy

Multi-Vector Attack

Mutual Authentication

National Institute of Standards and Technology (NIST)

National Vulnerability Database (NVD)

Nation-State Actor

Nested Virtualization

Network Access Control (NAC)

Network Address Hijacking

Network Address Translation (NAT)

Network Anomaly Detection

Network Architecture

Network Behavior Analysis (NBA)

Network Firewall

Network Flow Analysis

Network Forensics

Network Hardening

Network intrusion

Network Intrusion Detection System (NIDS)

Network Intrusion Prevention System (NIPS)

Network Mapper (Nmap)

Network Perimeter

Network Reconnaissance

Network Security

Network Security Monitoring

Network Segmentation

Network spoofing

Network Tap

Network Time Protocol (NTP)

NFC Attack

NIST Cybersecurity Framework (CSF)

NIST SP 800-53

Nonce

Non-Disclosure Agreement (NDA)

Non-repudiation

NoSQL Injection

Obfuscation

Object-Level Security

Offline Brute Force Attack

On-Demand Self-Service

One-Time Password (OTP)

Online Brute Force Attack

Open Port

Open Proxy

Open Redirect

Open Source Intelligence (OSINT)

Open Source Security

Open Source Security Testing Methodology Manual (OSSTMM)

Open Source Vulnerability Database (OSVDB)

Open Systems Interconnection (OSI) Model

Open Web Application Security Project (OWASP)

Operational audit

Operational Resilience

Operational Security (OPSEC)

Operational Technology (OT)

Outbound Firewall

Out-of-Band (OOB) Authentication

Out-of-Band Attack

Overflow Attack

Over-the-Air (OTA) Update

Over-the-Shoulder Attack

Packet Analyzer

Packet Capture

Packet Filtering

Packet Injection

Packet Loss

Packet Sniffer

Parameter Pollution

Passive Reconnaissance

Password Aging

Password Authentication Protocol (PAP)

Password Brute Forcing

Password Complexity

Password Cracking

Password Expiration

Password Guessing

Password Hash

Password History

Password Length

Password Manager

Password Policy

Password Reset

Password Spraying

Password Vault

Patch

Patch Management

Patch Tuesday

Payload

Payment Card Industry Data Security Standard (PCI DSS)

Penetration Testing (Pen Test)

Perimeter Defense

Perimeter Network

Perimeter Security

Personally Identifiable Information (PII)

Personally Identifiable Information (PII) Breach

Pharming

Phishing

Phreaking

Physical Access Control

Physical Security

Physical security breaches

Physical Security Information Management (PSIM)

Piggybacking

PII SCANNING

PING FLOOD

PING OF DEATH

PING SWEEP

PIVOTING

PLAINTEXT

PLAINTEXT ATTACK

PLATFORM AS A SERVICE (PAAS)

PLUG-AND-PLAY ATTACK

POINT-OF-SALE (POS) MALWARE

POINT-TO-POINT PROTOCOL (PPP)

POISONED NULL BYTE

POLICY-BASED ACCESS CONTROL (PBAC)

POLYMORPHIC MALWARE

PORT

PORT FORWARDING

PORT HONEYPOT

PORT KNOCKING

PORT MIRRORING

PORT SCANNING

PORT SECURITY

PORT TRIGGERING

PRETEXTING

PRETTY GOOD PRIVACY (PGP)

PRINTER HACKING

PRIVACY

PRIVACY BY DESIGN

PRIVACY IMPACT ASSESSMENT (PIA)

PRIVATE BRANCH EXCHANGE (PBX) HACKING

Private Key

Privilege

Privilege Creep

Privilege Escalation

Processor Security

Protected Health Information (PHI)

Protocol

Protocol Analysis

Protocol Analyzer

Proxy

Proxy Server

Pseudonymization

Pseudorandom Number Generator (PRNG)

Public Key

Public Key Cryptography

Public Key Cryptography Standards (PKCS)

Public Key Infrastructure (PKI)

Puffer

Purple Team

Push Button Reset Attack

Pwned

QRishing

Quantum Computing

Quantum Cryptography

Quantum cryptography attacks

Quarantine

Quarantine Network

Query Parameterization

Query String

RFID SKIMMING

RISK

RISK ACCEPTANCE

RISK APPETITE

RISK ASSESSMENT

RISK AVOIDANCE

RISK CULTURE

RISK GOVERNANCE

RISK HEAT MAP

RISK IDENTIFICATION

RISK MANAGEMENT

RISK MITIGATION

RISK REGISTER

RISK TOLERANCE

RISK TRANSFERENCE

RISK-BASED AUTHENTICATION

ROGUE ACCESS POINT

ROGUE ANTIVIRUS

ROGUE DHCP SERVER

ROGUE SOFTWARE

ROOT ACCESS

ROOT CERTIFICATE

ROOTING

ROOTKIT

ROOTKIT SCANNER

ROUTER

ROWHAMMER ATTACK

RUBBER DUCKY

RUNTIME APPLICATION SELF-PROTECTION (RASP)

Safe Browsing

Safe Harbor Principles

Salt

Salted Hash

Sandbox

Sandbox Evasion

Sarbanes-Oxley Act (SOX)

SCADA

Scalability

Scanning

Scareware

Screen Scraping

Script Kiddie

Secure Access Service Edge (SASE)

Secure Coding

Secure Copy Protocol (SCP)

Secure Electronic Transaction (SET)

Secure Enclave

Secure File Transfer Protocol (SFTP)

Secure Hash Algorithm (SHA)

Secure Multipurpose Internet Mail Extensions (S/MIME)

Secure Real-time Transport Protocol (SRTP)

Secure Shell (SSH)

Secure Socket Tunneling Protocol (SSTP)

Secure Sockets Layer (SSL)

Secure Software Development Lifecycle (SDLC)

Secure Web Gateway (SWG)

Security Assertion Markup Language (SAML)

Security Assessment

Security Audit

Security Automation

Security Awareness

Security Awareness Training

Security Baseline

Security Benchmark

Security Breach

Security Bulletin

Security Certification

Security Clearance

Security Compliance

Security Configuration Management (SCM)

Security Consulting

Security Control

Security Convergence

Security Engineering

Security Event

Security Event Log

Security Framework

Security Governance

Security Hardening

Security Incident

Security Incident Response

Security Information and Event Management (SIEM)

Security Kernel

Security Labeling

Security Misconfiguration

Security Models

Security Operations Center (SOC)

Security Orchestration

Security Orchestration, Automation, and Response (SOAR)

Security Patch

Security Policy

Security Posture

Security Procedures

Security Scorecard

Security Standard

Security Testing

Security Through Obscurity

Security Token

Security Token Service (STS)

Self-XSS

Sender Policy Framework (SPF)

Sensitive Data / Information

Sensitive Personal Information (SPI)

Server-Side Attack

Server-Side Request Forgery (SSRF)

Service Set Identifier (SSID)

Session cloning

Session Cookie

Session Fixation

Session Hijacking

Session Initiation Protocol (SIP)

Session Management

Session Prediction

Session replay

Shadow IT

Shared Responsibility Model

SHELL SHOCK

SHELLCODE

SHIMMING

SHORT MESSAGE SERVICE (SMS)

SHOULDER SURFING

SIDE CHANNEL ATTACK

SIDEJACKING

SIGNATURE-BASED DETECTION

SILVER TICKET ATTACK

SIMPLE MAIL TRANSFER PROTOCOL (SMTP)

SINGLE SIGN-ON (SSO)

SINKHOLE

SINKHOLE ROUTING

SKIMMING

SMART CARD

SMISHING

SMURF ATTACK

SNIFFER

SNORT

SOC 2

SOCIAL ENGINEERING ATTACK

SOCIAL ENGINEERING TOOLKIT (SET)

SOFTWARE AS A SERVICE (SAAS)

SOFTWARE BILL OF MATERIALS (SBOM)

SOFTWARE COMPOSITION ANALYSIS (SCA)

SOFTWARE DEFINED PERIMETER (SDP)

SOFTWARE DEVELOPMENT LIFE CYCLE (SDLC)

SOFTWARE EXPLOITATION

SOFTWARE VULNERABILITY

SUPPLY CHAIN RISK MANAGEMENT (SCRM)

SWATTING

SYMMETRIC ENCRYPTION

SYN FLOOD ATTACK

SYSLOG

SYSTEM BACKDOORS

SYSTEM HARDENING

SYSTEM INTEGRITY CHECK

SYSTEM LOGGING

TAILGATING

TARGETED ATTACK

TCP HIJACKING

TCP RST

TCP SYN

TECHNICAL VULNERABILITY

TEMPEST

THIN CLIENT

THIRD-PARTY COOKIE

THIRD-PARTY RISK MANAGEMENT (TPRM)

THREAT

THREAT ACTOR

THREAT ACTOR ATTRIBUTION

THREAT ASSESSMENT

THREAT EMULATION

THREAT HUNTING

THREAT INTELLIGENCE

THREAT INTELLIGENCE PLATFORM (TIP)

THREAT INTELLIGENCE SHARING

THREAT LANDSCAPE

Threat Modeling

Threat Surface

Threat Vector

Three-Way Handshake

Time-Based Blind SQL Injection

Time-Based One-Time Password (TOTP)

Time-of-Check to Time-of-Use (TOCTOU)

TimThumb Attack

Token

Tokenization

Traffic Analysis

Traffic Interception

Traffic Shaping

Transmission Control Protocol/Internet Protocol (TCP/IP)

Transport Layer

Transport Layer Security (TLS)

Transposition Cipher

Trapdoor Function

Triple DES (3DES)

Trojan / Trojan Horse

Trojan Downloader / Dropper

Trust Model

Trusted Execution Environment (TEE)

Trusted Platform Module (TPM)

Tunnel

Two-Factor Authentication (2FA) / Two-Step Verification

Typosquatting

Unified Extensible Firmware Interface (UEFI)

Unified Threat Management (UTM)

UNIFORM RESOURCE IDENTIFIER (URI)

UNIFORM RESOURCE LOCATOR (URL)

UNINTERRUPTIBLE POWER SUPPLY (UPS)

UNPATCHED VULNERABILITY

UNSOLICITED BULK EMAIL (UBE)

USA PATRIOT ACT

USER ACCOUNT CONTROL (UAC)

USER ACCOUNT CONTROL (UAC) BYPASS

USER BEHAVIOR ANALYTICS (UBA)

USER DATAGRAM PROTOCOL (UDP)

USER ENTITY AND BEHAVIOR ANALYTICS (UEBA)

USER PROVISIONING

VARIANT

VENDOR RISK MANAGEMENT (VRM)

VIRTUAL DESKTOP INFRASTRUCTURE (VDI)

VIRTUAL LOCAL AREA NETWORK (VLAN)

VIRTUAL MACHINE (VM) ESCAPE

VIRTUAL PRIVATE CLOUD (VPC)

VIRTUAL PRIVATE NETWORK (VPN)

VIRUS

VIRUS BULLETIN

VIRUS HOAX

VIRUS SCANNER

VIRUS SIGNATURE

VIRUS TOTAL

VOICE OVER IP (VOIP)

VOICE PHISHING (VISHING)

VULNERABILITY

VULNERABILITY ASSESSMENT

VULNERABILITY DATABASE

VULNERABILITY DISCLOSURE

VULNERABILITY LIFE CYCLE

VULNERABILITY MANAGEMENT

VULNERABILITY REMEDIATION

VULNERABILITY SCANNER

WAF BYPASS

WAR DIALING

WAR DRIVING

WAR ROOM

WARM SITE

WATERING HOLE ATTACK

WATERMARKING

WEAK CIPHER

WEAK PASSWORD

WEB APPLICATION FIREWALL (WAF)

WEB APPLICATION SECURITY

WEB BEACON

WEB BROWSER SECURITY

WEB CONTENT FILTERING

WEB CRAWLER

WEB DEFACEMENT

WEB EXPLOIT

WEB EXPLOITATION FRAMEWORK

WEB FILTERING

WEB PROXY

WEB PROXY AUTO-DISCOVERY PROTOCOL (WPAD)

WEB SCRAPING

WEB SERVER HARDENING

Web Server Log

Web Server Security

Web Services Security

Web Shell

Whaling

Whistleblower policy

White Box Testing

White Hat Hacker

Whitelist

Wide Area Network

Wi-Fi eavesdropping

Wi-Fi Protected Access (WPA)

Wi-Fi Protected Access II (WPA2)

Wi-Fi Protected Access III (WPA3)

WiFi Protected Setup (WPS)

Wi-Fi Security

Wildcard Certificate

Windows Management Instrumentation (WMI)

Wired Equivalent Privacy (WEP)

Wireless Access Point (WAP)

Wireless Intrusion Detection System (WIDS)

Wireless Intrusion Prevention System (WIPS)

Wireless LAN Controller (WLC)

Wireless Local Area Network (WLAN)

Wireless Security

Wireless Security Protocol (WSP)

Wireshark

Wiretapping

Worm

WPA/WPA2 CRACKING

X.509 CERTIFICATE

XMAS ATTACK

XML BOMB

XML ENCRYPTION

XML EXTERNAL ENTITY (XXE)

XML SIGNATURE

ZERO TRUST

ZERO TRUST ARCHITECTURE

ZERO TRUST NETWORK ACCESS (ZTNA)

ZERO-DAY ATTACK EXPLOIT

ZERO-KNOWLEDGE PROOF

ZOMBIE

About the author

With over twenty-six years in cybersecurity, Tolga TAVLAS knows his stuff. He's a leader in fintech and banking and is known for building secure digital solutions. He makes complicated cybersecurity topics easy to understand and shares his practical experience, making him a trusted expert in the field.

Tolga wrote "**CYBERSECURITY DICTIONARY for Everyone**" with the aim of making cybersecurity simple and accessible. This book breaks down complex terms into clear, easy-to-follow explanations, helping individuals and businesses protect their digital lives. Tolga aims to make cybersecurity a basic part of everyday digital knowledge.

Made in United States
Troutdale, OR
12/17/2024